DeWALT®

ELECTRICAL LICENSING

EXAM GUIDE

H. Ray Holder

Created exclusively
for DeWALT by:

PAL
publications®

wv   VMS, Inc.   :om
ALLEGAN MI 49010
www.VMS-online.com
1-800-343-6430

ISBN 0-9770003-4-6

Pal Publications, Inc.
374 Circle of Progress
Pottstown, PA 19464-3810

Copyright © 2005 by Pal Publications
First edition published 2005

Titles Available From DEWALT:

DEWALT Trade Reference Series

Construction Professional Reference

Datacom Professional Reference

Electric Motor Professional Reference

Electrical Estimating Professional Reference

Electrical Professional Reference

HVAC Professional Reference

Lighting & Maintenance Professional Reference

Plumbing Professional Reference

Referencia profesional sobre la industria eléctrica

Security, Sound & Video Professional Reference

Wiring Diagrams Professional Reference

DEWALT Exam and Certification Series

Electrical Licensing Exam Guide

HVAC Technician Certification Exam Guide

CONTENTS

DEWALT ELECTRICAL LICENSING EXAM GUIDE

by
H. Ray Holder

Notes

HOW TO PREPARE FOR THE EXAM

This book is a guide to preparing for the electricians' exams. It will not make you a competent electrician nor teach you the electrical trade, but it will give you an idea of the type of questions asked on most electricians' exams and how to answer them correctly.

Most exams consist of multiple-choice questions and these are the type reflected in this exam guide. They will give you a feel for how the examinations are structured and are based on many questions the author has encountered while taking numerous exams in recent years.

Begin your pre-exam preparation with two points in mind.

- OPPORTUNITIES in life will ARISE – be prepared for them.

- The more you LEARN – the more you EARN.

Attempting to take an exam without preparation is a complete waste of time. Attend classes at your local community college. Attend seminars, electrical code updates and company sponsored programs. Many major electrical suppliers and local unions sponsor classes of this type at no cost. Take advantage of them.

Become familiar with the National Electrical Code®; the Code has a LANGUAGE all its own. Understanding this language will help you to better interpret the NEC.® Do not become intimidated by its length. Become thoroughly familiar with the definitions in Chapter One; if you don't, the remainder of the NEC® will be difficult to comprehend. Remember, on the job we use different "lingo" and phrases compared to the way the NEC® is written and to the way many test questions are expressed.

HOW TO STUDY

Before beginning to study, get into the right frame of mind and relax. Study in a quiet place that is conducive to learning. If such a place is not available, go to your local library. It is important that you have the right atmosphere in which to study.

It is much better to study many short lengths of time than attempt to study fewer, longer lengths of time. Try to study about an hour every evening. You will need the support and understanding of your family to set aside this much needed time.

As you study this exam preparation book, the NEC® and other references, always highlight the important points. This makes it easier to locate Code references when taking the exam.

Use a straight edge such as a six-inch ruler when using the NEC® tables and charts. A very common mistake is to get on the wrong line when using these tables; when that happens, the result is an incorrect answer.

Use tabs on the major sections of your NEC® so they are faster and easier to locate when taking the exam. The national average allowed per question is less than three minutes; you cannot waste time.

WHAT TO STUDY

A common reason for one to be unsuccessful when attempting to pass electrical exams is not knowing what to study. Approximately forty percent of most exams are made up of "core" questions. These type of questions are reflected in this exam preparation book.

The subject matter covered in most electrical license examinations is:

- Grounding and bonding
- Overcurrent protection
- Wiring methods and installation
- Boxes and fittings
- Box and raceway fill
- Services and equipment
- Motors
- Special occupancies
- Load calculations
- Lighting
- Appliances
- Hazardous locations

Become very familiar with questions on the above. Knowing what to study is a major step toward passing your exam.

HELPFUL HINTS ON TAKING THE EXAM

- **Complete the easy questions first.** On most tests, all questions are valued the same. If you become too frustrated on any one question, it may have a negative effect upon your entire test.

- **Keep track of time.** Do not spend too much time on one question. If a question is difficult for you, mark the answer you think is correct and place a check (✔) by that question in the

Notes

Notes

examination booklet. Then go on to the next question. If you have time after finishing the rest of the exam, you can go back to the questions you have checked off. If you simply do not know the answer to a question, take a guess. Always choose the answer that is most familiar to you.

- **Only change answers if you know you are right.** Usually, your first answer is your best answer.

- **Relax.** Do not get uptight and stressed out when testing.

- **Tab your Code Book.** References are easier and faster to find.

- **Use a straight-edge.** Prevent getting on the wrong line when referring to the tables in the NEC.®

- **Get a good night's rest before the exam.** Do not attempt to drive several hours to an exam site; be rested and alert.

- **Understand the question.** One key word in a question can make a difference in what the question is asking. Underlining key words will help you to understand the meaning of the question.

- **Use a dependable calculator.** Use a solar-powered calculator that has a battery back-up. Since many test sites are not well lighted, this type of calculator will prepare you for such a situation. If possible, bring along a spare calculator.

- **Show up at least 30 minutes prior to your exam time.** Be sure to allow yourself time for traffic, etc. when planning your route to the exam location.

REGULATIONS AT THE EXAMINATION SITE

To ensure that all examinees are tested under equally favorable conditions, the following regulations and procedures are observed at most examination sites:

- Each examinee must present proper photo identification, preferably your driver's license, before you will be permitted to take the examination.

- No cameras, notes, tape recorders, pagers or cellular phones are allowed in the examination room.

- No one will be permitted to work beyond the established time limits.

- Examinees are not permitted any reference material EXCEPT the National Electrical Code.®

- Examinees will be permitted to use noiseless calculators during the examination. Calculators which provide programmable ability or pre-programmed calculators are prohibited.

- Permission of an examination proctor must be obtained before leaving the room while the examination is in progress.

TYPICAL EXAMINATION QUESTIONS

The following examples are intended to illustrate typical questions that appear on most electricians' exams.

EXAMPLE 1

An equipment grounding conductor of a branch circuit shall be identified by which of the following colors?

    a.   gray

    b.   white

    c.   black

    d.   green

Here you are asked to select from the listed colors the one that is to be used to identify the equipment grounding conductor of a branch circuit. Since Section 250.119 of the NEC® requires that green or green with yellow stripes be the color of insulation used on a grounding conductor (when it is not bare), the answer is (d).

EXAMPLE 2

A circuit leading to a gasoline dispensing pump must have a disconnecting means _____.

    a.   only in the grounded conductors.

    b.   only in the ungrounded conductors.

    c.   operating independently in all conductors.

    d.   that simultaneously disconnects both the grounded and ungrounded conductors feeding the dispensing pump.

Here the "question" is in the form of an incomplete statement. Your task is to select the choice that best completes the statement. In this case, you should have selected (d) since Section 514.11 of the NEC® specifies that such a circuit shall be provided with a means to disconnect simultaneously from the source of supply all conductors of a circuit, including the grounded conductor.

EXAMPLE 3

A building or other structure served shall be supplied by only one service EXCEPT one where the capacity requirements are in excess of _____.

    a.   800 amperes at a supply voltage of 600 volts or less.

    b.   1000 amperes at a supply voltage of 600 volts or less.

    c.   1500 amperes at a supply voltage of 600 volts or less.

    d.   2000 amperes at a supply voltage of 600 volts or less.

Again, the "question" is in the form of an incomplete statement and your task is to select the choice that best completes the statement. In this case, you are to find an exception. You have to select the condition that has to be met when supplying a building or structure by more than one service. You should have selected (d) because Section 230.2(C)(1) requires the conditions listed in (d) but does not require or permit the conditions listed in (a), (b) or (c).

EXAMPLE 4

Disregarding exceptions, the MINIMUM size overhead service-drop conductor shall be _____ AWG copper.

    a.   6

    b.   8

    c.   12

    d.   14

Here the "question" is in the form of a fill in the blank and your task is to select the choice that best completes the statement. In this case, exceptions are not applicable. You have to select the minimum size conductor required for overhead service-drop conductors. You should have selected (b) because Section 230.23(B) specifies that the conductors shall not be smaller than size 8 AWG copper.

HOW TO USE THIS BOOK

Each practice exam contained in this book consists of 25 questions. The time allotted for each practice exam is 75 minutes or 3 minutes per question. The final exams vary in time, length and difficulty depending upon the exam level. Using this time limit as a standard, you should be able to complete an actual examination in the allotted time.

To get the most out of this book you should answer every question and highlight your NEC® for future reference. If you have difficulty with a question and cannot come up with an answer that is familiar to you, put a check mark next to the question and come back to it after completing the remainder of the questions. Review your answers with the **ANSWER KEY** located in the back of this book. This will help you identify your strengths and weaknesses. When you discover you are weaker in some areas than others, you will know that further study is necessary in those areas.

Do only one practice exam contained in this book during an allotted study period. This way you do not get burned out and fatigued trying to study for too long a period of time. This also helps you develop good study habits.

GOOD LUCK!

Notes

HELPFUL REFERENCES

OHMS LAW CIRCLE FORMULAS

Ohms Law Circle

I=E/R
To Find Current

E=IxR
To Find Voltage

R=E/I
To Find Resistance

POWER CIRCLE FORMULAS

Power Circle

I=P/E
To Find Current

P=IxE
To Find Power

E=P/I
To Find Voltage

OHMS LAW/POWER
COMBINATION CIRCLE FORMULAS

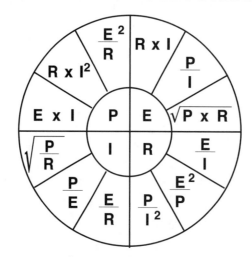

P = Power = Watts or Volt-Amperes
R = Resistance = Ohms
I = Current = Amperes
E = Force = Volts

TYPES OF POWER

Reactive power is supplied to a reactive load (capacitor/coil) and is measured in volt-amperes reactive (VAR). The capacitor on a motor uses reactive power to keep the capacitor charged. The capacitor uses no true power because it performs no actual work such as producing heat or motion.

In an AC circuit containing only resistance, the power in the circuit is true power. However, almost all AC circuits include capacitive reactance (capacitors) and/or inductive reactance (coils). Inductive reactance is the most common because all motors, transformers, solenoids and coils have inductive reactance.

Apparent power represents a load or circuit that includes both true power and reactive power and is expressed in volt-amperes (VA), kilovolt-amperes (kVA) or megavolt-amperes (MVA). Apparent power is a measure of component or system capacity because apparent power considers circuit current regardless of how it is used. For this reason, transformers are sized in volt-amperes rather than in watts.

TRUE POWER AND APPARENT POWER

True power is the actual power used in an electrical circuit and is expressed in watts (W). Apparent power is the product of voltage and current in a circuit calculated without considering the phase shift that may be present between total voltage and current in the circuit. Apparent power is measured in volt-amperes (VA).

True power equals apparent power in an electrical circuit containing only resistance. True power is less than apparent power in a circuit containing inductance or capacitance. A phase shift exists in most AC circuits that contain devices causing capacitance or inductance.

Capacitance is the property of a device that permits the storage of electrically separated charges when potential differences exist between the conductors. Inductance is the property of a circuit that causes it to oppose a change in current due to energy stored in a magnetic field; i.e., coils.

To calculate true power, apply the formula:

$$P_T = (I)^2 \times R$$

where

P_T = true power (in watts)

I = total circuit current (in amperes)

R = total resistive component of the circuit (in ohms)

To calculate apparent power, apply the formula:

$$P_A = E \times I$$

where

P_A = apparent power (in volt-amperes)

E = measured voltage (in volts)

I = measured current (in amperes)

POWER FACTOR FORMULA

Power factor is the ratio of true power used in an AC circuit to apparent power delivered to the circuit.

$$PF = \frac{P_T}{P_A}$$

where

PF = power factor (percentage)

P_T = true power (in watts)

P_A = apparent power (in volt-amperes)

USEFUL FORMULAS

To Find	Single Phase	Three Phase	Direct Current
Amperes when kVA is known	$\dfrac{\text{kVA} \times 1000}{E}$	$\dfrac{\text{kVA} \times 1000}{E \times 1.732}$	not applicable
Amperes when Horsepower is known	$\dfrac{\text{HP} \times 746}{E \times \%\text{EFF} \times \text{PF}}$	$\dfrac{\text{HP} \times 746}{E \times 1.732 \times \%\text{EFF} \times \text{PF}}$	$\dfrac{\text{HP} \times 746}{E \times \%\text{EFF}}$
Amperes when Kilowatts are known	$\dfrac{\text{kW} \times 1000}{E \times \text{PF}}$	$\dfrac{\text{kW} \times 1000}{E \times 1.732 \times \text{PF}}$	$\dfrac{\text{kW} \times 1000}{E}$
Kilowatts	$\dfrac{I \times E \times \text{PF}}{1000}$	$\dfrac{I \times E \times 1.732 \times \text{PF}}{1000}$	$\dfrac{I \times E}{1000}$
Kilovolt-Amperes	$\dfrac{I \times E}{1000}$	$\dfrac{I \times E \times 1.732}{1000}$	not applicable
Horsepower	$\dfrac{I \times E \times \%\text{EFF} \times \text{PF}}{746}$	$\dfrac{I \times E \times 1.732 \times \%\text{EFF} \times \text{PF}}{746}$	$\dfrac{I \times E \times \%\text{EFF}}{746}$
Watts	$E \times I \times \text{PF}$	$E \times I \times 1.732 \times \text{PF}$	$E \times I$

I = Amperes
E = Volts
kW = Kilowatts
kVA = Kilovolt-Amperes

HP = Horsepower
%EFF = Percent Efficiency
PF = Power Factor

VOLTAGE DROP FORMULAS

The NEC® recommends a maximum 3% voltage drop for either the branch circuit or the feeder.

Single Phase:

$$VD = \frac{2 \times K \times I \times D}{CM}$$

Three Phase:

$$VD = \frac{1.732 \times K \times I \times D}{CM}$$

VD = Volts (voltage drop of the circuit)

K = 12.9 Ohms/Copper or 21.2 Ohms/Aluminum (resistance constants for a conductor that is 1 foot long and 1 circular mil in diameter at an operating temperature of 75°C.)

I = Amperes (load at 100 percent)

D = Distance in feet (length of circuit from load to power supply)

CM = Circular-Mils (conductor wire size)

2 = Single-Phase Constant

1.732 = Three-Phase Constant

CONDUCTOR LENGTH/VOLTAGE DROP

Voltage drop can be reduced by limiting the length of the conductors.

Single Phase:

$$D = \frac{CM \times VD}{2 \times K \times I}$$

Three Phase:

$$D = \frac{CM \times VD}{1.732 \times K \times I}$$

CONDUCTOR SIZE/VOLTAGE DROP

Increase the size of the conductor to decrease the voltage drop of the circuit (reduce its resistance).

Single Phase:

$$CM = \frac{2 \times K \times I \times D}{VD}$$

Three Phase:

$$CM = \frac{1.732 \times K \times I \times D}{VD}$$

MAINTENANCE ELECTRICIAN
Practice Exam #1

The following questions are based on the 2005 edition of the National Electrical Code® and are typical of questions encountered on most Maintenance Electricians' Exams. Select the best answer from the choices given and review your answers with the answer key included in this book.

ALLOTTED TIME: 75 minutes

Maintenance Electrician
Practice Exam #1

1. The National Electrical Code is NOT _____.

    a. designed for future expansion of electrical use

    b. designed to safeguard people and property from electrical hazards

    c. published by the NFPA

    d. intended as a specification manual for trained persons

2. The NEC mandates specific branch circuits, receptacle outlets and utilization equipment to be provided with a ground-fault circuit interrupter (GFCI); this device is intended _____.

    a. to prevent overloading the conductors

    b. to prevent overloading the circuit breakers

    c. for the protection of equipment from ground-faults

    d. for the protection of personnel

3. Electrical wiring installed _____ is considered to be installed in a damp location.

    a. under canopies or roofed open porches

    b. underground

    c. outside

    d. none of these apply

4. When a 20 ampere, 120 volt receptacle outlet is installed, in which of the following listed locations is the receptacle required to have GFCI protection?

    a. in a classroom of an educational facility

    b. in the lobby of a movie theater

    c. under an outdoor canopy, near the entrance of a public library

    d. in the walkway of an enclosed retail shopping mall

5. Which of the following is a unit of electrical power?

    a. watt
    b. voltage
    c. resistance
    d. conductance

6. When checking for continuity between a circuit breaker and the neutral bar of a panelboard when using a continuity tester, positive continuity is indicated. The reason may be _____.

    I. a conductor grounded

    II. a light fixture or an appliance may be turned on

    a. I only
    b. II only
    c. either I or II
    d. neither I nor II

7. When installing electrical metallic tubing (EMT), there shall be no more than _____ 90° bends in the tubing between boxes and/or pull points.

    a. 3
    b. 4
    c. 5
    d. 6

8. The full-load running current of a 3 HP, 208 volt, single-phase, continuous duty, AC motor is _____.

    a. 10.6 amperes
    b. 19.6 amperes
    c. 13.2 amperes
    d. 18.7 amperes

Notes

Notes

9. When a 225 ampere-rated panelboard contains twenty (20) snap switches (circuit breakers) rated at 30 amperes, the panelboard shall have overcurrent protection not in excess of _____ amperes.

    a. 200

    b. 225

    c. 300

    d. 600

10. If the voltage drop of a branch circuit is too great, which of the following, if any, may be the result?

    I. unsatisfactory illumination of lighting fixtures

    II. overheating of a motor or unsatisfactory motor speed

    a. I only

    b. II only

    c. both I and II

    d. neither I nor II

11. What is the MINIMUM required distance from the floor to the ceiling in front of a 400 ampere-rated motor control center, disconnect switch or main switchboard?

    a. 6 feet

    b. 6 feet 3 inches

    c. 6 feet 6 inches

    d. 8 feet

12. The NEC mandates the largest size insulated solid conductor permitted to be installed into an existing raceway is _____.

    a. 4 AWG

    b. 6 AWG

    c. 8 AWG

    d. 10 AWG

13. A panelboard may contain no more than _____ branch-circuit overcurrent protection devices.

    a. 24

    b. 30

    c. 42

    d. 48

14. Which one of the following listed circuit breakers is NOT a standard ampere rating listed in the NEC?

    a. 110 ampere

    b. 90 ampere

    c. 75 ampere

    d. 225 ampere

15. A 4 inch x 1½ inch metal octagon box with a flat blank cover may contain a MAXIMUM of _____ size 14 AWG conductors.

    a. 6

    b. 7

    c. 9

    d. 10

16. In general, electrical nonmetallic tubing (ENT) shall be securely fastened at intervals not exceeding _____.

    a. 10 feet

    b. 6 feet

    c. 4 feet

    d. 3 feet

Notes

Notes

17. In general, when a 20 ampere-rated branch circuit supplies a single receptacle outlet, the rating of the receptacle must not be less than _____ amperes.

    a. 10

    b. 15

    c. 16

    d. 20

18. A branch circuit supplying a 7½ HP, 240 volt, single-phase, induction-type, continuous duty, Design B, AC motor shall have an ampacity of at least _____ amperes.

    a. 60

    b. 40

    c. 50

    d. 75

19. If the copper conductors supplying the previously referenced motor have a temperature rating of 75° C, determine the MINIMUM size as required by the NEC.

    a. 8 AWG

    b. 6 AWG

    c. 10 AWG

    d. 4 AWG

20. Which of the following fuses, if any, shall be used for replacement only in existing installations and not permitted for use in new installations?

    I. Edison-base

    II. Class H cartridge

    a. I only

    b. II only

    c. both I and II

    d. neither I nor II

21. A 240 volt, single-phase, 10 kW commercial dishwasher will have a full-load current rating of _____.

    a. 21 amperes

    b. 24 amperes

    c. 42 amperes

    d. 30 amperes

22. The ampacity of a conductor is defined by the NEC as being the current in amperes a conductor can carry continuously under the conditions of use without exceeding _____.

    a. its temperature rating

    b. the allowable voltage drop limitations

    c. its melting point

    d. its rated voltage

23. That portion of a wiring system that is between the final overcurrent device protecting the circuit and the outlets is defined as a _____.

    a. branch circuit

    b. feeder

    c. sub-feeder

    d. service drop

24. Given: a 120 volt branch circuit is to serve fifteen (15), 150 watt incandescent luminaires (light fixtures). Determine the current.

    a. 36.50 amperes

    b. 18.75 amperes

    c. 9.37 amperes

    d. 14.10 amperes

Notes

Notes

25. In general, all switches and circuit breakers used as switches shall be located so that they may be operated from a readily accessible place. They shall be installed so that the center of the grip of the operating handle of the switch or circuit breaker, when at its highest position, is NOT more than _____ above the floor or platform.

    a. 5½ feet

    b. 6 feet

    c. 6½ feet

    d. 6 feet 7 inches

MAINTENANCE ELECTRICIAN
Practice Exam #2

The following questions are based on the 2005 edition of the National Electrical Code® and are typical of questions encountered on most Maintenance Electricians' Exams. Select the best answer from the choices given and review your answers with the answer key included in this book.

ALLOTTED TIME: 75 minutes

Maintenance Electrician
Practice Exam #2

1. In computing the load of luminaires (lighting fixtures) that employ ballasts, transformers or auto-transformers such as fluorescent lighting units, the load must be based on the _____.

    a. wattage of the lamps

    b. VA of the lamps

    c. length of the lighting fixture

    d. ampere ratings of the complete units

2. A branch-circuit, inverse time circuit breaker rated in amperes shall be permitted as a controller for _____.

    a. continuous duty motors

    b. intermittent duty motors

    c. torque motors

    d. all motors

3. Compliance with the provisions of the NEC will result in an electrical installation that is essentially _____.

    a. free from hazard

    b. a good electrical system

    c. an efficient system

    d. all of these

4. Under normal conditions, an equipment grounding conductor will _____.

    a. improve current flow

    b. not carry current

    c. carry current

    d. reduce circuit resistance

5. Cord-and-plug-connected vending machines located in break rooms of commercial establishments are required to be provided with _____.

   I. a power supply cord not smaller than size 12 AWG with a length not exceeding 4 feet
   II. GFCI protection

   a. I only
   b. II only
   c. both I and II
   d. neither I nor II

6. A motor with nine leads coming out of it is a _____.

   a. two-phase motor
   b. two-speed motor
   c. dual-voltage motor
   d. multispeed motor

7. On a four-wire, delta-connected system, the conductor having the higher voltage to ground (high-leg) shall be identified as _____ in color if the grounded conductor is also present to supply lighting or similar loads.

   a. white
   b. red
   c. green
   d. orange

8. When an electrical trade size ¾ inch electrical metallic tubing (EMT) is in excess of 24 inches in length, the conduit is permitted to contain no more than _____ size 10 AWG conductors with THWN insulation.

   a. 6
   b. 10
   c. 8
   d. 14

Notes

Notes

9. Given: a 50 foot electrical metallic tubing (EMT) to be installed will contain a total of nine conductors. Six of these wires are considered to be current-carrying. What is the derating factor, in percent, that must be applied to the ampacity of the current-carrying conductors?

    a. 80 percent
    b. 70 percent
    c. 60 percent
    d. 50 percent

10. The reason the NEC requires all grounded and ungrounded conductors of a common circuit to be installed together in the same metal raceway is to reduce _____.

    a. expense
    b. inductive heat
    c. voltage drop
    d. resistance

11. In general, electrical metallic tubing (EMT) shall be securely fastened within ____ of each junction box, panelboard or other conduit termination.

    a. 3 feet
    b. 6 feet
    c. 8 feet
    d. 10 feet

12. One horsepower is equal to _____ watts.

    a. 746
    b. 1000
    c. 1250
    d. 1500

13. For continuous duty motors, the motor nameplate current rating is used to determine the size of the _____ required for the motor.

    a. disconnecting means

    b. branch-circuit conductors

    c. motor overload protection

    d. short-circuit protection

14. In general, branch-circuit conductors that supply a single AC motor used in a continuous duty application shall have an ampacity of NOT less than _____ percent of the motor's full-load running current rating.

    a. 125

    b. 100

    c. 150

    d. 115

15. Unless listed and identified otherwise, terminations and conductors of circuits rated 100 amperes or larger shall be rated at _____.

    a. 60° C

    b. 75° C

    c. 90° C

    d. 120° C

16. When circuit breakers are used to switch 120 volt and 277 volt, high-intensity discharge, lighting branch circuits, the circuit breakers shall be listed and marked _____.

    I. SWD

    II. HID

    a. I only

    b. II only

    c. either I or II

    d. neither I nor II

Notes

17. An insulated conductor with a continuous yellow-colored insulation is considered to be a/an _____.

    a. equipment grounding conductor

    b. grounded conductor

    c. switch-leg

    d. ungrounded conductor

18. When an insulated black conductor is identified by three continuous white stripes along its entire length, it is considered a/an _____.

    a. grounded conductor

    b. ungrounded conductor

    c. equipment grounding conductor

    d. phase of a delta-connected system

19. Given: a 9 kW rated, 208 volt, three-phase electric steamer is to be installed in the kitchen of a commercial establishment. The steamer will draw _____ amperes.

    a. 43.0

    b. 4.3

    c. 25.0

    d. 75.0

20. What is the approximate MAXIMUM distance a single-phase, 240 volt, 42 ampere load may be located from a panelboard given the following related information?

    – for copper conductors, K = 12.9

    – size 8 AWG THWN/THHN conductors are used

    – limit voltage drop to 3%

    a. 50 feet

    b. 110 feet

    c. 160 feet

    d. 195 feet

21. The NEC permits electrical trade size ⅜ inch flexible metal conduit (FMC) to be used for tap conductors (fixture whips) to luminaires (lighting fixtures) provided the length of the flex does not exceed _____.

  a.   4 feet
  b.   6 feet
  c.   8 feet
  d. 10 feet

22. The MINIMUM required width of the working space in front of panelboards, switchboards, disconnects and motor controllers shall be the width of the equipment or _____, whichever is greater.

  a.   2 feet
  b.   3 feet
  c.   4 feet
  d. 30 inches

23. In general, branch-circuit conductors serving continuous loads such as luminaires (lighting fixtures) in an office building or commercial establishment shall have an ampacity of NOT less than _____ percent of the load.

  a. 125
  b. 115
  c. 150
  d.   80

24. Transient voltage surge suppressors (TVSS) may be installed on circuits of up to _____ volts.

  a.   150
  b.   300
  c.   600
  d. 1000

Notes

25. In general, the overcurrent protection for size 12 AWG copper conductors, regardless of the insulation type, shall not exceed _____ amperes.

    a. 15

    b. 20

    c. 25

    d. 30

RESIDENTIAL ELECTRICIAN
Practice Exam #3

The following questions are based on the 2005 edition of the National Electrical Code® and are typical of questions encountered on most Residential Electricians' Exams. Select the best answer from the choices given and review your answers with the answer key included in this book.

ALLOTTED TIME: 75 minutes

Residential Electrician
Practice Exam #3

1. Metal enclosures for grounding electrode conductors shall be _____.

    a. electrically continuous

    b. electrically isolated

    c. not permitted

    d. rigid metal conduit (RMC) only

2. Unless listed for the control of other loads, general-use dimmer switches shall only be used to control _____.

    a. switched receptacles for cord-connected incandescent luminaires (lighting fixtures)

    b. permanently installed incandescent luminaires (lighting fixtures)

    c. a ceiling fan with luminaire (lighting fixture)

    d. holiday decorative lighting

3. Which of the following is a false statement?

    a. Receptacle outlets are required on decks and balconies of multi-family dwellings.

    b. The required outdoor receptacle outlet installed at multi-family dwellings must be no higher than 6½ feet above grade.

    c. A receptacle outlet is required at each unit of a multi-family dwelling that has an individual ground level entrance.

    d. The required outdoor receptacle outlet for multi-family dwellings may be at the front or back of the unit.

4. The MAXIMUM allowable ampacity of size 18 AWG fixture wire is _____ amperes.

    a. 12

    b. 10

    c. 8

    d. 6

5. A receptacle outlet installed for a gas or electric clothes dryer in the laundry room of a dwelling unit must be installed within at least _____ feet of the intended location of the dryer.

    a. 6
    b. 4
    c. 10
    d. 3

6. Which of the following listed conductors is required to have overcurrent protection on a residential electric service?

    a. grounding conductor
    b. bonding conductor
    c. identified conductors
    d. ungrounded conductors

7. Given: a 40 gallon electric water heater has a nameplate rating of 4,500 watts @ 240 volts. What is the MAXIMUM standard size circuit breaker the NEC permits to protect the water heater?

    a. 20 amperes
    b. 25 amperes
    c. 30 amperes
    d. 45 amperes

8. If a cable or raceway enters an enclosure (such as a meter base) above uninsulated live parts in a wet location, the fittings used to connect the enclosure must be _____ for wet locations.

    a. identified
    b. marked
    c. suitable
    d. listed

Notes

9.   The power supply cord to a mobile home shall have a MAXIMUM rating of _____.

    a.   50 amperes

    b.   60 amperes

    c.   100 amperes

    d.   125 amperes

10.   A 150 kVA, single-phase transformer having a secondary voltage of 120/240 has been installed at a multi-family dwelling. The current available at the secondary is _____ amperes.

    a.   329

    b.   421

    c.   625

    d.   729

11.   When the heating, air-conditioning or refrigeration equipment is installed on the roof of an apartment complex, a 15 or 20 ampere-rated receptacle outlet _____.

    a.   is not required by the NEC

    b.   may be connected to the line side of the equipment disconnecting means if the outlet is GFCI protected

    c.   shall be located on the same level and within 25 feet of the heating, air-conditioning or refrigeration equipment

    d.   shall be installed on the roof where the equipment is located and not more than 75 feet from each unit

12.   The MINIMUM number of 120 volt, 15 ampere, lighting branch circuits required for a dwelling that has 70 feet by 30 feet of habitable space is _____.

    a.   2

    b.   3

    c.   4

    d.   5

13. Receptacle outlets on a 20 ampere-rated, small-appliance branch circuit serving kitchen countertops in a dwelling unit are to be rated _____.

   I.   15 amperes
   II. 20 amperes

      a. I only
      b. II only
      c. either I or II
      d. neither I nor II

14. In general, service conductors installed as open conductors shall have a clearance of not less than _____ from doors, porches, balconies or windows that are designed to be opened.

      a. 6 feet
      b. 8 feet
      c. 5 feet
      d. 3 feet

15. Given: a 120 volt lighting branch circuit has only six (6), 100 watt, 120 volt, incandescent lighting fixtures connected to it. Determine the total current on this branch circuit.

      a. 30 amperes
      b. 20 amperes
      c.   5 amperes
      d. 12 amperes

16. Raceways installed in solid rock shall be permitted to be buried at lesser depths than required when covered by at least _____ of concrete extending down to the rock.

      a. 2 inches
      b. 4 inches
      c. 6 inches
      d. 8 inches

Notes

17. Electrical outlet boxes installed in walls and ceilings are required to be _____.

    a. visible

    b. readily accessible

    c. accessible

    d. metal

18. The MINIMUM internal depth of a box that is intended to enclose a flush-mounted device shall be _____.

    a. ½ inch

    b. 9/16 inch

    c. 15/16 inch

    d. 1¼ inches

19. Metallic surface-type cabinets for electrical equipment located in damp or wet locations shall be mounted so there is at LEAST _____ air space between the cabinet and the wall or other supporting surface.

    a. ⅛ inch

    b. ¼ inch

    c. ⅜ inch

    d. ½ inch

20. In dwelling units, hallways of _____ feet or more in length are required to have at least one (1), 15 or 20 ampere-rated receptacle outlet.

    a. 10

    b. 12

    c. 15

    d. 20

21. Receptacles in a kitchen of a residence that are to serve countertop surfaces shall be installed so that no point along the wall line is more than _____ inches measured horizontally from a receptacle outlet in that space.

    a. 24

    b. 18

    c. 36

    d. 48

22. Given: you are to install a 20 ampere, 120 volt, GFCI-protected, buried, UF cable branch circuit that serves a yard light for a residence. The UF cable is required to have a ground cover of at LEAST _____.

    a. 6 inches

    b. 12 inches

    c. 18 inches

    d. 24 inches

23. Where NM cable is run at angles with joists in unfinished basements of dwellings, it shall be permissible to secure cables not smaller than _____ AWG directly to the lower edges of the joists.

    a. 10/2

    b. 8/2

    c. 6/2

    d. 6/3

24. When exceptions are not to be taken into consideration, the ungrounded conductor of an electrical branch circuit may be identified by _____.

    a. white or gray-colored insulation

    b. bare or green-colored insulation

    c. only gray-colored insulation

    d. black or red-colored insulation

Notes

Notes

25. Three-way and four-way switches shall be so wired that all switching is done_____.

     a. only in the grounded circuit conductor

     b. only in the ungrounded circuit conductor

     c. either in the grounded or ungrounded circuit conductor

     d. only in the white circuit conductor

RESIDENTIAL ELECTRICIAN
Practice Exam #4

The following questions are based on the 2005 edition of the National Electrical Code® and are typical of questions encountered on most Residential Electricians' Exams. Select the best answer from the choices given and review your answers with the answer key included in this book.

ALLOTTED TIME: 75 minutes

Residential Electrician
Practice Exam #4

1. The MAXIMUM height that kitchen countertop receptacle outlets may be installed above the countertop shall be _____.

    a. 12 inches

    b. 18 inches

    c. 20 inches

    d. 24 inches

2. For dwelling units, panelboards are permitted to be located in all the areas listed EXCEPT _____.

    a. hallways

    b. garages

    c. kitchens

    d. bathrooms

3. The MAXIMUM number of grounded conductors allowed to be terminated in an individual terminal, if conductors are not parallelled, in the neutral bus of a panelboard is _____.

    a. 1

    b. 2

    c. 3

    d. 4

4. Flexible cord is considered as protected by a 20 ampere branch circuit breaker if it is _____ and approved for use with a specific listed appliance.

    a. not less than 6 feet in length

    b. size 20 AWG or larger

    c. size 18 AWG or larger

    d. size 16 AWG or larger

5. A 120 volt receptacle is installed on a dwelling unit kitchen island countertop that is 8 feet from the kitchen sink. Which one of the following statements, if any, is correct?

    a. GFCI protection is not required because the receptacle is not within 6 feet of the sink.

    b. GFCI protection is required for all countertop kitchen receptacles.

    c. GFCI protection is not required on receptacles installed on kitchen islands.

    d. None of the above is correct.

6. When installing replacement luminaires (lighting fixtures) in outlets that do not provide an equipment grounding conductor, which of the following is/are permitted?

    a. Install a listed luminaire and add GFCI protection to the branch circuit.

    b. Install a bonding jumper between a metal lighting fixture and a metal box.

    c. Install a listed luminaire which completely covers the outlet box.

    d. None of the above is correct.

7. A GFCI-protected receptacle outlet that provides power to a swimming pool recirculating pump motor shall be permitted NOT closer than _____ feet from the inside wall of the pool.

    a. 5

    b. 4

    c. 12

    d. 15

8. The grounding contacts of branch-circuit receptacle outlets shall be grounded by connection to the _____ conductor.

    a. bonding

    b. neutral

    c. grounded

    d. equipment grounding

Notes

Notes

9. Any one cord-and-plug-connected utilization equipment connected to a 20 ampere-rated branch circuit shall have a MAXIMUM rating of _____.

    a. 10 amperes
    b. 16 amperes
    c. 20 amperes
    d. 25 amperes

10. In walls or ceilings with a surface of concrete, tile, gypsum, plaster or other noncombustible material, boxes employing a flush-type cover or faceplate shall be installed so that the front edge of the box will not set back of the finished surface more than _____.

    a. ¼ inch
    b. ½ inch
    c. ⅜ inch
    d. ¾ inch

11. For a one-family dwelling, the service disconnecting means shall have a rating of not less than _____ amperes when supplied with a 120/240 volt, single-phase service drop.

    a. 30
    b. 60
    c. 100
    d. 200

12. What is the feeder or service demand load, in kW, for one (1), 8 kW rated, residential electric range?

    a. 8.0 kW
    b. 6.4 kW
    c. 7.0 kW
    d. 12.0 kW

13. Mobile home service equipment shall be rated NOT less than _____ at 120/240 volt, single-phase.

    a.   50 amperes

    b.   60 amperes

    c.   100 amperes

    d.   150 amperes

14. Underwater luminaires (fixtures) installed in swimming pools are required to have GFCI protection if they operate at a voltage greater than _____.

    a.   15 volts

    b.   50 volts

    c.   120 volts

    d.   150 volts

15. Given: the ungrounded service-entrance conductors for a single-family residence are size 3/0 AWG copper conductors. A copper grounding electrode conductor attached to a water pipe electrode shall not be smaller than size _____.

    a.   2 AWG

    b.   4 AWG

    c.   6 AWG

    d.   8 AWG

16. What is the MAXIMUM time period allowed for outdoor Christmas decoration lighting for residences?

    a.   30 days

    b.   60 days

    c.   90 days

    d.   four months

Notes

/9j/4AAQSkZJRgABAQEAYABgAAD/2wBDAAGVlZk
unused placeholder

21. The NEC mandates the ampacity of UF cable shall be that of
    _____ conductors.

    a. 60° C rated

    b. 75° C rated

    c. 86° F rated

    d. 90° C rated

22. The front edge of a switch box installed in a wall constructed of
    wood shall be _____ from the surface of the wall.

    a. flush with or projected out

    b. set back a maximum of ¼ inch

    c. set back a maximum of ½ inch

    d. set back a maximum of ⅜ inch

23. For dwelling units where receptacle outlets are installed within at
    least _____ of the outside edge of a sink in a laundry room, GFCI
    protection must be provided.

    a. 4 feet

    b. 6 feet

    c. 8 feet

    d. 10 feet

24. For dwelling units, a branch circuit supplying _____ receptacle
    outlets is permitted to also supply receptacle outlets in an
    attached garage.

    a. outdoor

    b. bathroom

    c. laundry room

    d. kitchen small-appliance

Notes

Notes

25. A metal underground water pipe may serve as a grounding electrode if it is in direct contact with the earth for at least _____ feet or more.

a. 6

b. 8

c. 10

d. 12

RESIDENTIAL ELECTRICIAN
Practice Exam #5

The following questions are based on the 2005 edition of the National Electrical Code® and are typical of questions encountered on most Residential Electricians' Exams. Select the best answer from the choices given and review your answers with the answer key included in this book.

ALLOTTED TIME: 75 minutes

Residential Electrician
Practice Exam #5

1. What is the MAXIMUM allowable cord length for the cord supplying a cord-and-plug-connected dishwasher installed under a kitchen counter in a dwelling unit?

    a. 1½ feet

    b. 2 feet

    c. 3 feet

    d. 4 feet

2. What is the MINIMUM size copper SE cable with type THHW insulation that may be used as ungrounded, service-entrance phase conductors for a 150 ampere-rated, 120/240 volt, single-phase residential service?

    a. 1/0 AWG

    b. 1 AWG

    c. 2 AWG

    d. 3 AWG

3. For outlet boxes designed to support ceiling-suspended (paddle) fans that weigh more than _____ pounds, the required marking shall include the maximum weight to be supported.

    a. 20

    b. 30

    c. 35

    d. 70

4. A connection to a driven or buried grounding electrode shall _____.

    a. be accessible

    b. not be required to be accessible

    c. not be permitted to be buried

    d. be visible

5. What is the MAXIMUM allowable standard circuit breaker rating that may be used with a size 8 AWG, copper, NM-B cable supplying an electric range?

    a. 30 amperes

    b. 40 amperes

    c. 50 amperes

    d. 60 amperes

6. The MAXIMUM number of size 10 AWG conductors permitted in a 4 inch x 1¼ inch octagon metal junction box is _____.

    a. 2

    b. 4

    c. 5

    d. 6

7. Determine the MINIMUM number of 15 ampere, 120 volt, lighting branch circuits required for a dwelling that has 2,600 square feet of habitable space.

    a. 3

    b. 4

    c. 5

    d. 6

8. When calculating the total load on a dwelling unit, how many VA per square feet must be included for the general-purpose receptacle outlets?

    a. 0

    b. 1

    c. 2

    d. 3

Notes

Notes

9. When calculating the total load for a dwelling unit, what is the MINIMUM load, in VA, that must be added for the two (2), required, small-appliance branch circuits?

    a. 1,200 VA

    c. 2,400 VA

    b. 1,500 VA

    d. 3,000 VA

10. It shall be permissible to apply a demand factor of _____ percent to the nameplate rating load of four (4) or more fastened-in-place electric water heaters in a multi-family dwelling unit.

    a. 50

    b. 75

    c. 80

    d. 90

11. In dwelling units, at least one (1) wall receptacle shall be installed in bathrooms; the receptacle outlet shall be within at LEAST _____ of the outside edge of each basin.

    a. 12 inches

    b. 18 inches

    c. 24 inches

    d. 36 inches

12. When installing NM cable through bored holes in wooden studs, the holes shall be bored so that the edge of the hole is not less than _____ from the edge, or the cable shall be protected by a steel plate at least ¹⁄₁₆ inch thick.

    a. ¾ inch

    b. 1 inch

    c. 1¼ inches

    d. 1½ inches

13. A 240 volt, single-phase, 30 ampere-rated, electric clothes dryer will have a VA rating of _____.

    a. 5,000

    b. 7,200

    c. 8,000

    d. 15,400

14. Type XHHW insulated conductors are permitted to be used in _____.

    a. dry locations only

    b. wet locations only

    c. dry or damp locations only

    d. dry, damp or wet locations

15. When two (2) single-pole switches are mounted on the same strap, the number of conductors permitted in the box shall be reduced by _____ conductor(s).

    a. 1

    b. 2

    c. 3

    d. 4

16. When doing residential service and feeder calculations, electric clothes dryers are to be calculated at a MINIMUM of _____ watts (VA) or the nameplate rating, whichever is larger.

    a. 3,000

    b. 4,500

    c. 5,000

    d. 6,000

Notes

Notes

17. The rating of any single cord-and-plug-connected appliance used on a 30 ampere-rated branch circuit shall not exceed _____.

     a. 30 amperes

     b. 27 amperes

     c. 24 amperes

     d. 16 amperes

18. In general, all receptacles installed on 15 and 20 ampere, 120 volt branch circuits are required to _____.

     a. be GFCI protected

     b. have a nonmetallic faceplate

     c. be of the ungrounded type

     d. be of the grounding type

19. A branch circuit supplying a 5 kW wall-mounted oven and a 7 kW counter-mounted cook-top in a residence has a demand factor of _____ on the ungrounded service-entrance conductors.

     a. 12.0 kW

     b. 9.5 kW

     c. 8.0 kW

     d. 7.8 kW

20. The MINIMUM size copper equipment grounding conductor required to equipment served by a 40 ampere-rated branch circuit is _____.

     a. 10 AWG

     b. 8 AWG

     c. 12 AWG

     d. 14 AWG

21. What is the MAXIMUM allowable voltage between conductors on a branch circuit supplying luminaires in a residence?

    a. 120 volts

    b. 150 volts

    c. 240 volts

    d. 250 volts

22. The ampacity of the branch-circuit conductors to a residential, central-heating electric furnace shall NOT be less than _____ percent of the furnace load.

    a. 80

    b. 100

    c. 115

    d. 125

23. Switches and circuit breakers used as a disconnecting means to appliances shall be _____.

    a. factory installed only

    b. of the fusible type

    c. of the indicating type

    d. of the tripping type

24. The NEC requires recessed portions of fixture enclosures that are not identified for contact with insulation to be spaced from combustible material a MINIMUM of _____.

    a. ³⁄₈ inch

    b. ¹⁄₂ inch

    c. ³⁄₄ inch

    d. 1 inch

Notes

Notes

25. When installing luminaires (lighting fixtures) without GFCI protection above an indoor spa or hot tub, they are required to be mounted at LEAST _____ above the maximum water level of the spa or hot tub.

    a.   5 feet

    b.   7 feet 6 inches

    c.   6 feet 7 inches

    d. 12 feet

RESIDENTIAL ELECTRICIAN
Practice Exam #6

The following questions are based on the 2005 edition of the National Electrical Code® and are typical of questions encountered on most Residential Electricians' Exams. Select the best answer from the choices given and review your answers with the answer key included in this book.

ALLOTTED TIME: 75 minutes

Residential Electrician
Practice Exam #6

1. When a single branch circuit with an equipment grounding conductor is feeding a detached garage from a house, what, if any, additional grounding electrode(s) is/are required at the garage?

    a. A ground rod and underground metal water pipe.

    b. A metal underground water pipe only.

    c. A metal underground water pipe and building steel, if available.

    d. No additional electrodes are required.

2. The point of attachment of a service drop to a residence where the voltage is 120 volts to ground is required to be at LEAST _____ above grade.

    a. 8 feet

    b. 10 feet

    c. 12 feet

    d. 15 feet

3. The branch-circuit overcurrent protection device shall be permitted to serve as the disconnecting means for stationary motors rated NOT greater than _____ horsepower.

    a. 2

    b. 1

    c. ½

    d. ⅛

4. When installing an overhead service using a rigid metal conduit mast for the support of the service-drop conductors, the mast shall be of adequate strength or be _____.

    a. at least 2 inches in diameter

    b. a minimum of 3 inches in diameter

    c. supported by braces or guys

    d. less than 4 feet in length

5. When wall switches are installed in the same room with a spa or hot tub, the MINIMUM distance from the switches to the inside wall of the hot tub or spa shall be _____.

    a.   5 feet

    b.   10 feet

    c.   15 feet

    d.   18 feet

6. All 15 and 20 ampere, single-phase, 125 volt receptacle outlets located at LEAST _____ from the inside walls of a swimming pool shall be provided with GFCI protection.

    a.   10 feet

    b.   15 feet

    c.   20 feet

    d.   25 feet

7. A luminaire (lighting fixture) shall be supported independently of the outlet box when it weighs more than _____ pounds, unless the outlet box is listed for the weight to be supported.

    a.   6

    b.   25

    c.   35

    d.   50

8. Of the following listed connectors, which type is prohibited for connection of a grounding conductor to an enclosure?

    a.   sheet metal screw

    b.   pressure connector

    c.   clamp

    d.   lug

Notes

9. A ground rod is required to be driven a MINIMUM of _____ feet into the soil.

    a. 4
    b. 6
    c. 8
    d. 10

10. Receptacle outlets are permitted to be installed directly over a bathtub or shower stall _____.

    a. if the outlet has a weatherproof cover
    b. if the outlet is GFCI protected
    c. if the outlet has a weatherproof cover and is GFCI protected
    d. never

11. Receptacle outlets installed in the _____ position in countertops are prohibited.

    a. vertical
    b. horizontal
    c. face-up
    d. all of these are acceptable

12. Receptacle outlets in a kitchen of a residence that are to serve countertop surfaces shall be installed so that no point along the wall line is more than _____ when measured horizontal from a receptacle in that space.

    a. 24 inches
    b. 18 inches
    c. 36 inches
    d. 48 inches

13. The ampacity of branch-circuit conductors and the rating or setting of overcurrent protective devices supplying fixed, outdoor, electric deicing equipment shall not be less than _____ percent of the full-load current of the equipment.

    a. 100

    b. 125

    c. 80

    d. 150

14. When an intermediate metal conduit (IMC) contains conductors of 600 volts or less and is buried under a gravel driveway of a single-family dwelling, the IMC must be buried at LEAST _____.

    a. 12 inches

    b. 18 inches

    c. 24 inches

    d. 30 inches

15. Disregarding exceptions, where residential lighting outlets are installed in interior stairways, there shall be a wall switch provided _____.

    a. near the stairs

    b. every seven steps

    c. at the top and bottom of the stairs if there are more than six steps

    d. at any convenient location

16. Where it is impractical to locate the service head or gooseneck above the point of attachment of the service-drop conductors, the service head or gooseneck shall be permitted no further than _____ from the point of attachment.

    a. 3 feet

    b. 4 feet

    c. 6 feet

    d. 2 feet

Notes

17. A metal box or terminal fitting having separately bushed holes for each conductor shall be used whenever change is made from conduit to _____.

    a. knob-and-tube wiring

    b. nonmetallic-sheathed cable

    c. type AC cable

    d. type MC cable

18. Electric water heaters for swimming pools shall have the heating elements subdivided into loads not exceeding _____.

    a. 32 amperes

    b. 36 amperes

    c. 48 amperes

    d. 60 amperes

19. Concealed knob-and-tube wiring shall be permitted to be used only for _____.

    a. dwellings units

    b. extensions of existing installations

    c. accessible installations

    d. installing temporary wiring

20. In dwelling units, receptacle outlets installed within 6 feet of the outside edge of a sink in a laundry room shall _____.

    a. have a weatherproof cover

    b. have AFCI protection

    c. be a single receptacle

    d. be provided with GFCI protection

21. The largest size copper conductor permitted for nonmetallic-sheathed cable is size _____AWG.

    a. 4

    b. 2

    c. 6

    d. 1

22. For the purpose of determining box fill, size 12 AWG copper conductors with THHN insulation are to be calculated at _____.

    a. 1.75 cubic inches

    b. 2.00 cubic inches

    c. 2.25 cubic inches

    d. 2.50 cubic inches

23. What is the MAXIMUM distance allowed between supports when installing type NM cable?

    a. 3 feet

    b. 4½ feet

    c. 6 feet

    d. 10 feet

24. Disregarding exceptions, the MINIMUM size underground service-lateral conductors permitted by the NEC is _____.

    a. 8 AWG copper

    b. 6 AWG copper

    c. 4 AWG aluminum

    d. 2 AWG aluminum

Notes

Notes

25. How many size 12/2 with ground, type NM cables are permitted to be installed in an outlet box with a total volume of 18 cubic inches and with a duplex receptacle outlet in the box?

    a. 1

    b. 2

    c. 3

    d. 4

JOURNEYMAN ELECTRICIAN
Practice Exam #7

The following questions are based on the 2005 edition of the National Electrical Code® and are typical of questions encountered on most Journeyman Electricians' Exams. Select the best answer from the choices given and review your answers with the answer key included in this book.

ALLOTTED TIME: 75 minutes

Journeyman Electrician
Practice Exam #7

1. A listed instantaneous water heater shall be permitted to be subdivided into circuits protected at NOT more than _____.

    a. 60 amperes

    b. 100 amperes

    c. 120 amperes

    d. 150 amperes

2. Individual ungrounded branch-circuit conductors must be separately identified if _____.

    a. they are accessible

    b. they are in the same raceway or enclosure with ungrounded conductors from a different system

    c. they are part of a premises wiring system with branch-circuit conductors from a different system

    d. all of these apply

3. The continuity of the grounding conductor system for portable electrical carnival equipment shall be verified ____.

    a. and recorded on an annual basis

    b. and recorded on a quarterly basis

    c. and recorded on a monthly basis

    d. each time the equipment is connected

4. A plate-type grounding electrode shall be installed below the surface of the earth NOT less than _____.

    a. 24 inches

    b. 30 inches

    c. 36 inches

    d. 48 inches

5. Determine the MAXIMUM allowable ampacity of a size 1/0 AWG, THW, copper, current-carrying conductor installed in a common raceway with three (3) other current-carrying conductors of the same size. Ambient temperature is 86° F.

    a. 150 amperes

    b. 105 amperes

    c. 112 amperes

    d. 120 amperes

6. The NEC defines a continuous load to be a load where the maximum current on the circuit is expected to continue for _____ hours or more.

    a. 3

    b. 4

    c. 6

    d. 8

7. Determine the MINIMUM size THW, copper, branch-circuit conductors required by the NEC to supply a three-phase, continuous duty, AC motor that has a full-load current of 70 amperes. Assume all terminations to be rated at 75° C.

    a. 1 AWG

    b. 2 AWG

    c. 3 AWG

    d. 4 AWG

8. Given: you have determined a conductor has a computed ampacity of 75 amperes. What is the MAXIMUM standard ampere rating of the overcurrent protection device the NEC permits to protect this circuit? This is not a motor circuit or part of a multi-outlet branch circuit supplying receptacles.

    a. 70 amperes

    b. 75 amperes

    c. 80 amperes

    d. 85 amperes

Notes

9. A 208 volt, three-phase, 50 HP, squirrel-cage, continuous duty, Design C motor has a full-load running current of _____.

    a. 130 amperes

    b. 143 amperes

    c. 162 amperes

    d. 195 amperes

10. Which of the following metal piping systems contained in a building is/are required to be bonded?

    a. A gas pipe for a gas heater.

    b. A bathroom water pipe connected to an electric water heater.

    c. An air line for compressed air.

    d. All of these need to be bonded.

11. Determine the MINIMUM trade size electrical metallic tubing (EMT) required to enclose eight (8), size 6 AWG copper conductors with THHW insulation installed in a 60 foot conduit run.

    a. 1 inch

    b. 1¼ inch

    c. 1½ inch

    d. 2 inch

12. Given: a feeder supplies a continuous load of 240 amperes. In general, the overcurrent protection device protecting this circuit shall have a MINIMUM rating of _____.

    a. 240 amperes

    b. 300 amperes

    c. 250 amperes

    d. 275 amperes

13. Liquidtight flexible metal conduit (LFMC) shall be securely fastened within _____ of each box or other conduit termination.

    a.   4½ feet

    b.   3 feet

    c.   12 inches

    d.   18 inches

14. When a pull box contains conductors of size 4 AWG and larger and a straight pull of the conductors is to be made, the length of the box shall not be less than ____ times the trade diameter of the largest conduit entering the box.

    a.   six

    b.   four

    c.   eight

    d.   twelve

15. In general, all of the following wiring methods are permitted to be used as a feeder for temporary wiring on a construction site EXCEPT _____.

    a.   NM cable

    b.   NMC cable

    c.   type PD cord

    d.   type SJO cord

16. Determine the MAXIMUM number of size 14 AWG THHN conductors permitted to be installed in a trade size ⅜ inch flexible metal conduit (FMC) that contains a bare size 14 AWG grounding conductor when the flex has external fittings.

    a.   2

    b.   3

    c.   4

    d.   5

Notes

Notes

17. Which of the following listed conduits does the NEC permit to enclose conductors feeding wet-niche underwater pool lights?

    a. electrical metallic tubing (EMT)

    b. electrical nonmetallic tubing (ENT)

    c. galvanized rigid metal conduit (RMC)

    d. schedule 40 rigid nonmetallic conduit (RNC)

18. Underground wiring is not permitted under a swimming pool or within _____ of the pool unless the wiring is supplying pool equipment.

    a. 10 feet

    b. 6 feet

    c. 5 feet

    d. 8 feet

19. Which of the following listed types of batteries is/are NOT permitted for use as a source of power for emergency systems?

    a. automotive type

    b. lead-acid type

    c. alkali type

    d. all of these

20. The MINIMUM rating for a service disconnecting means for a building, other than a one-family dwelling, consisting of three (3), two-wire circuits shall be _____.

    a. 15 amperes

    b. 30 amperes

    c. 60 amperes

    d. 100 amperes

21. In guest rooms of hotels and motels, at LEAST _____ general-purpose receptacle(s) installed in the room is/are required to be readily accessible.

    a. 1

    b. 2

    c. 3

    d. all

22. When branch-circuit conductors are contained inside a ballast compartment and they are within 3 inches of the ballast, the conductors shall have a temperature rating of at LEAST _____.

    a. 105° C

    b. 90° C

    c. 75° C

    d. 60° C

23. When a size 4 AWG or larger conductor enters a panelboard, which of the following must be provided?

    a. a bonding jumper

    b. a grounding clip

    c. an insulated bushing

    d. an insulated grounding conductor

24. Class _____ locations are those that are hazardous because of the presence of easily ignitible fibers or flyings.

    a. I

    b. II

    c. III

    d. IV

Notes

Notes

25. A single-phase, 240 volt, 15 kVA generator may be loaded to no more than _____ amperes per phase.

    a. 62.50

    b. 31.25

    c. 41.66

    d. 52.50

JOURNEYMAN ELECTRICIAN
Practice Exam #8

The following questions are based on the 2005 edition of the National Electrical Code® and are typical of questions encountered on most Journeyman Electricians' Exams. Select the best answer from the choices given and review your answers with the answer key included in this book.

ALLOTTED TIME: 75 minutes

Notes

Journeyman Electrician
Practice Exam #8

1. Motor control circuits shall be arranged so that they will be disconnected from all sources of supply when _____.

    a. the disconnecting means is in the open position

    b. the disconnecting means is not within sight of the controller

    c. the disconnecting means is integral with the controller

    d. installed in a motor control center

2. Which of the following is permitted for closing an unused circuit breaker opening in panelboards?

    a. Installation of a closure plate or other device that fits securely and closes the opening.

    b. Installation of a spare or unused circuit breaker that fills the opening and is listed for the panelboard.

    c. Installation of a closure plate listed for the purpose by the manufacturer.

    d. All of these are permitted.

3. Before demand factors are taken into consideration for commercial buildings, general-purpose receptacle loads are to be computed at not less than _____ VA per outlet.

    a. 100

    b. 120

    c. 150

    d. 180

4. Which of the following listed conductor insulations has a greater temperature rating when used in a dry location compared to when used in a wet location?

    a. THW

    b. RHW

    c. THWN

    d. THHW

5. Which of the following situations regarding grounding would NOT be in compliance with the standards set forth by the NEC?

    I.   Steel reinforcing bars to be poured in a concrete foundation and not connected as part of the grounding system.

    II. Concrete foundations with steel reinforcing bars which were not available for use as a grounding electrode because the concrete was poured before the electrical contractor arrived on the job.

    a. I only
    b. II only
    c. both I and II
    d. neither I nor II

6. In a health care facility, low-voltage electrical equipment that is likely to become energized and is frequently in contact with the bodies of patients shall operate on a voltage of _____ volts or less if the equipment is not approved as intrinsically safe, double-insulated or moisture-resistant.

    a. 10
    b. 24
    c. 100
    d. 120

7. When ambient temperature is not a factor, a size 8 AWG single copper conductor with type FEPB insulation installed in free air will have a MAXIMUM allowable ampacity of ____.

    a. 80 amperes
    b. 55 amperes
    c. 45 amperes
    d. 83 amperes

8. For the purpose of determining conductor fill in a device box, a switch is counted as equal to _____ conductor(s), based on the largest conductor connected to the switch.

    a. 0

    b. 1

    c. 2

    d. 3

9. When determining the required ampacity of the branch-circuit conductors to supply an AC, three-phase, continuous duty motor, current values of the motor given on _____ shall be used.

    a. the motor's nameplate

    b. Table 430.248

    c. Table 430.52

    d. Table 430.250

10. When calculating the total load for a commercial building, what is the MINIMUM computed load, in volt-amps, required by the NEC for a branch circuit supplying an exterior sign?

    a. 1,200 volt-amperes

    b. 1,800 volt-amperes

    c. 1,500 volt-amperes

    d. 2,000 volt-amperes

11. Determine the MAXIMUM number of size 1 XHHW compact conductors permitted in a trade size 3 inch, Schedule 40, rigid nonmetallic conduit (RNC) that is 100 feet long.

    a. 21

    b. 19

    c. 14

    d. 18

12. The MINIMUM headroom of working space about service equipment and motor control centers shall be at least _____.

    a. 6 feet

    b. 6 feet 6 inches

    c. 6 feet 7 inches

    d. 8 feet

13. Class _____ hazardous locations are those in which flammable gases or vapors are or may be present in the air in quantities sufficient to produce explosive or ignitible mixtures.

    a. I

    b. II

    c. III

    d. IV

14. In indoor areas where walls are frequently washed such as laundries or car washes, metal conduit and metal panelboards shall be mounted with a _____ space between the wall and conduit or panelboard when the equipment is installed exposed.

    a. ¾ inch

    b. ½ inch

    c. ⅛ inch

    d. ¼ inch

15. A branch circuit supplied by size 12 AWG conductors is a _____-rated branch circuit when protected by a 15 ampere-rated circuit breaker.

    a. 20 ampere

    b. 25 ampere

    c. 15 ampere

    d. 10 ampere

Notes

Notes

16. For other than single-family dwellings, spas and hot tubs are required to have a clearly labeled emergency shutoff or control switch provided. This switch shall be installed at a point readily accessible to the users and not less than _____ feet and within sight of the spa or hot tub.

    a. 12
    b. 10
    c. 15
    d. 5

17. In a data-processing room, the disconnecting means for the data-processing equipment shall be _____.

    a. within sight of the equipment
    b. near an exit door
    c. at the main disconnect
    d. within 30 feet of the equipment

18. Type MI cable shall be supported at intervals not exceeding _____.

    a. 2 feet
    b. 4 feet
    c. 6 feet
    d. 10 feet

19. The internal depth of an outlet box used only to splice conductors to a luminaire (light fixture) shall NOT be less than _____.

    a. ½ inch
    b. $^{15}/_{16}$ inch
    c. ¾ inch
    d. 1 inch

20. Direct-buried conductors emerging from grade shall be protected by raceways or covers up to at LEAST _____ above finished grade.

    a. 8 feet

    b. 10 feet

    c. 12 feet

    d. 15 feet

21. What is the MINIMUM size equipment grounding conductor required by the NEC for a branch circuit protected by a 50 ampere-rated circuit breaker?

    a. 12 AWG

    b. 10 AWG

    c. 8 AWG

    d. 6 AWG

22. Disregarding exceptions, all three-phase, 277/480 volt, wye electrical services require ground-fault protection for each service disconnecting means when rated at LEAST _____ or more.

    a. 400 amperes

    b. 600 amperes

    c. 1,000 amperes

    d. 1,500 amperes

23. Determine the conductor ampacity given the following:

    – conductors are size 500 kcmil copper

    – conductor insulation is THWN

    – eight (8) current-carrying conductors are in the raceway

    – ambient temperature is 125° F

    – conduit length is 75 feet

    – installation is a wet location

    a. 178.2 amperes

    b. 199.5 amperes

    c. 294.6 amperes

    d. 380.0 amperes

Notes

24. In general, the electrical supply equipment for an RV park back-in site shall be located on the left side of the parked vehicle not more than _____ from the left edge of the stand.

    a.   5 feet

    b.   7 feet

    c.   15 feet

    d.   30 feet

25. Given: a trade size ½ inch rigid metal conduit (RMC) is to be installed in a Class I location and a trade size ½ inch conduit seal is required. The MINIMUM thickness of the sealing compound shall not be less than _____.

    a.   ⅜ inch

    b.   ½ inch

    c.   ⅝ inch

    d.   ¾ inch

JOURNEYMAN ELECTRICIAN
Practice Exam #9

The following questions are based on the 2005 edition of the National Electrical Code® and are typical of questions encountered on most Journeyman Electricians' Exams. Select the best answer from the choices given and review your answers with the answer key included in this book.

ALLOTTED TIME: 75 minutes

Journeyman Electrician
Practice Exam #9

1. Where a water pipe passes over an indoor motor-control center, leak protection apparatus shall have a MINIMUM clearance of _____.

    a. 6 feet to the protection apparatus

    b. 6 feet to the water pipe

    c. 6½ feet to the protection apparatus

    d. 6½ feet to the water pipe

2. The MINIMUM size, single-conductor, medium-voltage cable that may be installed in a cable tray is _____.

    a. 1/0 AWG

    b. 2/0 AWG

    c. 3/0 AWG

    d. 4/0 AWG

3. The MAXIMUM standard rating of the initial nontime-delay fuse for the branch-circuit, short-circuit and ground-fault protection of a 1 HP, three-phase, 208 volt, squirrel-cage, AC motor shall be _____.

    a. 6 amperes

    b. 10 amperes

    c. 15 amperes

    d. 20 amperes

4. For Class III equipment that is not subject to overloading, the MAXIMUM surface temperature under operating conditions shall not exceed _____.

    a. 120° C

    b. 165° C

    c. 248° F

    d. 210° C

5. When a metal wireway contains splices and taps, the conductors, splices and taps shall not fill the wireway to more than _____of its area at that point.

    a. 20 percent

    b. 30 percent

    c. 40 percent

    d. 75 percent

6. In an underground rigid nonmetallic conduit (RNC) system that consists of 50 feet in length between pulling points, what is the MAXIMUM number of bends this run may have?

    a. 4 – 90° bends

    b. 6 – 90° bends

    c. 4 – 120° bends

    d. 2 – 90° bends

7. Grounding electrodes made of pipe or conduit shall not be smaller than _____ electrical trade size.

    a. ½ inch

    b. ¾ inch

    c. 1 inch

    d. 1½ inch

8. Liquidtight flexible metallic conduit (LFMC) shall NOT be used _____.

    a. in lengths in excess of 6 feet

    b. in concealed work

    c. in hazardous locations

    d. where subject to physical damage

Notes

Notes

9. Determine the allowable ampacity of a size 3 AWG, THHN, copper conductor given the following:

    – Two (2) current-carrying conductors are in the conduit.

    – The ambient temperature is 34° C.

    – The terminations are rated at 60° C.

        a. 105.6 amperes

        b.   85.0 amperes

        c.   81.6 amperes

        d. 110.0 amperes

10. When an overcurrent protective device for a transient voltage surge suppressor (TVSS) is located at the service location, it _____.

        a. is not to be counted as a service disconnect

        b. shall be considered as a service disconnecting means

        c. shall be more than 10 feet from the main disconnecting means

        d. shall be prohibited

11. Standard trade size 2 inch rigid nonmetallic conduit (RNC) shall be supported at least every _____.

        a. 3 feet

        b. 5 feet

        c. 6 feet

        d. 8 feet

12. When sizing time-delay Class CC fuses for motor branch-circuit, short-circuit and ground-fault protection, they are to be sized at the same value as _____.

        a. inverse time circuit breakers

        b. nontime-delay fuses

        c. instantaneous trip circuit breakers

        d. adjustable trip circuit breakers

13. Where sizes 1/0 AWG through 4/0 AWG single-conductor cables are installed in a ladder-type cable tray, the MAXIMUM allowable rung spacing for the ladder cable tray shall be_____.

    a. 6 inches

    b. 9 inches

    c. 12 inches

    d. 15 inches

14. The NEC requires the MAXIMUM length permitted for a flexible cord supplying a 208 volt, single-phase, room air conditioner to be _____.

    a. 4 feet

    b. 6 feet

    c. 8 feet

    d. 10 feet

15. In general, which of the following receptacle outlets in a commercial kitchen are required by the NEC to be GFCI protected?

    a. All 125 volt, single-phase, 15 and 20 ampere-rated receptacles.

    b. All 125 volt, 15 or 20 ampere-rated receptacles in wet locations only.

    c. All receptacle outlets.

    d. All 125 volt countertop receptacle outlets only.

16. All wiring installed in or under an aircraft hanger floor shall comply with the requirements for _____ locations.

    a. Class I, Division 1

    b. Class I, Division 2

    c. Class II, Division 1

    d. Class II, Division 2

Notes

17. Branch-circuit conductors supplying more than one motor shall have an ampacity of at least _____% of the FLC of the largest motor, and 100% of the FLC of the other motor(s) in the group.

  a. 25

  b. 80

  c. 100

  d. 125

18. What is the MINIMUM size equipment grounding conductor required for a 5 HP, three-phase, 208 volt motor having 20 ampere-rated overload protection and short-circuit overcurrent protection rated at 30 amperes?

  a. 10 AWG

  b. 14 AWG

  c. 12 AWG

  d. 8 AWG

19. If a circuit breaker serves as the controller for a motor and the motor is not in sight of the breaker, the NEC requires which of the following?

  a. The motor to be less than 2 HP.

  b. The breaker be able to be locked in the open position.

  c. The motor to be Code letter "E".

  d. The breaker to be rated 25,000 AIC.

20. In Class II, Division 1 hazardous locations, an approved method of connection of conduit to boxes is _____.

  a. compression fittings

  b. threaded bosses

  c. welding

  d. all of these

21. Given: an AC service is supplied with four parallel sets of size 500 kcmil aluminum conductors. What is the MINIMUM size copper grounding electrode required when connected to a metal underground water line used as the grounding electrode system?

    a. 3/0 AWG

    b. 4/0 AWG

    c. 250 kcmil

    d. 2/0 AWG

22. Given: an AC transformer arc welder has a 50 ampere-rated primary current and a 60 percent duty-cycle. Determine the MINIMUM size copper 60° C rated conductors the NEC requires to supply this welder.

    a. 6 AWG

    b. 8 AWG

    c. 10 AWG

    d. 4 AWG

23. When installing wiring for sensitive electronic equipment, the MAXIMUM voltage to ground is required to be _____ volts.

    a. 30

    b. 60

    c. 120

    d. 300

24. In a health care facility, a patient bed location in a critical care area is required to have which of the following?

    a. four single receptacles or two duplex receptacles

    b. six single receptacles or three duplex receptacles

    c. four duplex receptacles or eight single receptacles

    d. two single receptacles or one duplex receptacle

Notes

25. Which of the following circuits is prohibited for the grounded conductor to be dependent on receptacle devices for continuity?

   a.   all circuits

   b.   multi-outlet circuits

   c.   GFCI-protected circuits

   d.   multi-wire circuits

JOURNEYMAN ELECTRICIAN
Practice Exam #10

The following questions are based on the 2005 edition of the National Electrical Code® and are typical of questions encountered on most Journeyman Electricians' Exams. Select the best answer from the choices given and review your answers with the answer key included in this book.

ALLOTTED TIME: 75 minutes

**Journeyman Electrician
Practice Exam #10**

1. Where conduits enter a floor-standing switchboard or panelboard at the bottom, the conduits, including their end fittings, shall not rise more than _____ above the bottom of the enclosure.

    a. 6 inches

    b. 4 inches

    c. 2 inches

    d. 3 inches

2. When installed on the service of a building or structure, a transient voltage surge suppressor (TVSS) shall be connected _____.

    a. on the line side of the service disconnect overcurrent device

    b. on the load side of the service disconnect overcurrent device

    c. on the meter housing

    d. at the final overcurrent device

3. A detached garage fed from a dwelling with a three-wire feeder of size 4/0 AWG, aluminum, USE cable with a 200 ampere-rated circuit breaker for overcurrent protection requires a MINIMUM size copper grounding electrode to a concrete encased electrode of _____ AWG.

    a. 8

    b. 6

    c. 4

    d. 2

4. When a raceway is run from the interior to the exterior of a building, the raceway shall _____.

    a. be rigid metal conduit (RMC)

    b. be rigid nonmetallic conduit (RNC)

    c. be sealed

    d. have an explosionproof seal

5. A disconnecting means shall be provided on the supply side of all fuses in circuits having a voltage of at least _____ volts to ground.

    a. 120

    b. 250

    c. 150

    d. 277

6. Given: an 80 ampere, 240 volt, single-phase load is located 200 feet from the supply panelboard and is wired with size 3 AWG copper conductors with THHN/THWN insulation. What is the approximate voltage drop on this circuit? (K = 12.9)

    a. 6 volts

    b. 4 volts

    c. 9.2 volts

    d. 7.84 volts

7. Enclosures in a Class I, Division 1 location containing components that have arcing devices must have an approved seal located within _____ of each conduit run entering or leaving such enclosures.

    a. 12 inches

    b. 18 inches

    c. 24 inches

    d. 30 inches

8. Surface-mounted incandescent luminaires (light fixtures) installed on the wall above the door or on the ceiling of a clothes closet shall have a clearance of at LEAST _____ between the luminaire and the nearest point of storage.

    a. 6 inches

    b. 8 inches

    c. 18 inches

    d. 12 inches

Notes

Notes

9.   The NEC considers the area around the dispensing pumps at a service station to be a hazardous location. This area extends a height of 18 inches above grade and up to a distance from the enclosure of the dispensing pump of _____.

    a.   5 feet

    b.   10 feet

    c.   16 feet

    d.   20 feet

10. According to the NEC, what are the two mandatory branches of the hospital emergency system?

    a.   the emergency branch and the standby branch

    b.   the life safety branch and the critical branch

    c.   the normal branch and the alternate branch

    d.   the essential branch and the non-essential branch

11. Class 3 single-conductors shall not be smaller than _____.

    a.   18 AWG

    b.   16 AWG

    c.   14 AWG

    d.   20 AWG

12. Given: an office building has a 208Y/120 volt, three-phase service with a balanced net computed load of 90 kVA. What is the current each ungrounded conductor carries at full load?

    a.   188 amperes

    b.   250 amperes

    c.   433 amperes

    d.   750 amperes

13. When grounding the structural reinforcing steel of a swimming pool, what is the smallest size grounding conductor permitted for this installation?

   a. 12 AWG

   b. 10 AWG

   c.   8 AWG

   d.   6 AWG

14. When used as a service lateral, how far may type USE conductors extend into a building when the conductors terminate within an enclosure?

   a. 3 feet

   b. 6 feet

   c. Nearest the point of entrance.

   d. The conductors shall terminate outside of the building.

15. Storage batteries used as a source of power for emergency systems shall be of a suitable rating and capacity to supply and maintain the total load for at LEAST _____.

   a. ½ hour

   b. 1 hour

   c. 1½ hours

   d. 2 hours

16. Given: a size 14 AWG branch-circuit conductor protected by a 15 ampere-rated circuit breaker is to serve three (3), 20 ampere-rated duplex receptacles. This branch circuit _____.

   a. would be in compliance with the NEC if the wire was size 12 AWG

   b. would be in compliance with the NEC if the circuit breaker was rated at 20 amperes

   c. is not in compliance with the NEC because the receptacle outlets have a 20 ampere rating

   d. is in compliance with the NEC

Notes

17. Where fuses are used for motor overload protection and the supply system is a three-wire, three-phase AC with one conductor grounded, the NEC requires at LEAST _____ fuse(s).

    a. 0

    b. 1

    c. 2

    d. 3

18. Given: a metal junction box has 27 cubic inches of space and contains a total of six (6), size 12 AWG wires. Additional wires of size 10 AWG need to be added in the box. No grounding conductors, devices or fittings are contained in the box. What is the MAXIMUM number of size 10 AWG conductors that may be added to this box?

    a. 2

    b. 5

    c. 6

    d. 8

19. For enclosing a mobile home supply cord, what is the MAXIMUM trade size conduit permitted between the branch-circuit panelboard of a mobile home and the underside of a mobile home floor?

    a. 1 inch

    b. 1¼ inch

    c. 1½ inch

    d. 2 inch

20. Which of the following wiring methods is/are approved for use for fixed wiring in an area above Class I locations in a commercial garage?

    a. type MI cable

    b. type TC cable

    c. type MC cable

    d. all of the above

21. Class II locations are those that are hazardous (classified) because of the presence of _____.

    a. flammable gas

    b. ignitible fibers

    c. ignitible vapors

    d. combustible dust

22. Branch circuits supplying a fixed, storage-type, electric water heater with a capacity of 120 gallons or less shall NOT be less than _____ of the full-load current of the water heater.

    a. 80 percent

    b. 100 percent

    c. 125 percent

    d. 150 percent

23. Household-type appliances with surface heating elements having a MAXIMUM demand of more than _____ amperes shall have its power supply subdivided into two or more circuits.

    a. 60

    b. 50

    c. 40

    d. 30

24. When driving a ground rod and solid rock is encountered, the ground rod is permitted to be buried in a trench that is at least _____ deep.

    a. 3 feet

    b. 4 feet

    c. 24 inches

    d. 30 inches

Notes

25. Ten (10) copper THW conductors are to be installed in a 20 foot length of intermediate metal conduit (IMC); five (5) size 1 AWG and five (5) size 3 AWG. What is the MINIMUM allowable trade size IMC required to contain these conductors?

  a. 1½ inch

  b. 2 inch

  c. 2½ inch

  d. 4 inch

JOURNEYMAN ELECTRICIAN
Practice Exam #11

The following questions are based on the 2005 edition of the National Electrical Code® and are typical of questions encountered on most Journeyman Electricians' Exams. Select the best answer from the choices given and review your answers with the answer key included in this book.

ALLOTTED TIME: 75 minutes

Journeyman Electrician
Practice Exam #11

1. Unless listed and identified otherwise, terminations and conductors of sizes 14 AWG through 1 AWG shall be rated at _____.

   a. 75° C

   b. 75° F

   c. 90° C

   d. 60° C

2. The MINIMUM size copper equipment bonding jumper on the supply side of a service consisting of parallel copper phase conductors with a total circular mil area of 2000 kcmil shall be ____.

   a. 2/0 AWG

   b. 3/0 AWG

   c. 4/0 AWG

   d. 250 kcmil

3. Which of the following grounding electrodes is NOT permitted to be used as an effective ground-fault current path?

   a. the metal frame of a building or structure

   b. a metal underground water pipe in direct contact with the earth for 10 feet or more

   c. an electrode encased by at least 2 inches of concrete

   d. the earth

4. Primary overcurrent devices, impedance limiting means or other inherent protective means shall be permitted to protect a motor-control circuit transformer when the transformer is rated no more than _____ VA.

   a. 50

   b. 60

   c. 75

   d. 100

5. When installing electric space-heating cables in ceilings, which of the following statements is/are true?

    a. The heating cables are permitted to be installed over walls.

    b. The non-heating leads of the cables shall be at least 6 inches in length.

    c. The heating cables shall be kept free from contact with other electrically conductive surfaces.

    d. All of these statements are true.

6. Given: an unbroken length of size 12 AWG conductor is looped inside a junction box so that there is more than 12 inches of free conductor. The conductor is not spliced and does not terminate on a device. What volume, in cubic inches, must be counted for the looped conductor?

    a. 2.25 cubic inches

    b. 3.25 cubic inches

    c. 4.50 cubic inches

    d. 4.00 cubic inches

7. Nonmetallic-sheathed cable is permitted to be used _____.

    a. in buildings of type III, IV, and V construction

    b. as open runs in suspended ceilings of commercial buildings

    c. in buildings of type I and II construction

    d. in conduit installed underground

8. Given: a boat repair shop has several 120 volt, 20 ampere, general-purpose receptacle outlets installed at a height of 16 inches in a work area where gasoline may be handled and mechanical ventilation is not provided. The outlets are considered to be installed in a/an _____ location.

    a. Class I, Division 1

    b. Class I, Division 2

    c. Class II, Division 1

    d. Unclassified

Notes

Notes

9. A luminaire (lighting fixture) rated at 7 amperes requires a MINIMUM size _____ AWG fixture wire.

   a. 12
   b. 14
   c. 16
   d. 18

10. All receptacle outlets located in areas of pediatric wards of hospitals are required to _____.

   I. be listed tamper-resistant
   II. employ a listed tamper-resistant cover

   a. I only
   b. II only
   c. either I or II
   d. neither I nor II

11. An installation of 40 feet of electrical metallic tubing (EMT) will contain the following conductors:

   – Three (3) size 2 AWG with THWN insulation
   – Four (4) size 6 AWG with THWN insulation
   – Three (3) size 10 AWG with THWN insulation

   This installation requires a MINIMUM trade size _____ EMT.

   a. 1¼ inch
   b. 1½ inch
   c. 2 inch
   d. 2½ inch

12. Where UF cable is used as a substitute for Type NM cable, the conductor insulation is required to be rated at _____.

   a. 60° C
   b. 75° C
   c. 75° F
   d. 90° C

13. When armored cable (AC) is installed in thermal insulation, the ampacity of the cable shall be that of _____ rated conductors.

     a. 60° C

     b. 75° C

     c. 194° F

     d. 90° C

14. Receptacle outlets installed on a branch circuit rated at 30 amperes are permitted to have an ampere rating of _____.

     a. 15, 20 or 30 amperes

     b. 20 or 30 amperes

     c. 30 or 40 amperes

     d. 30 amperes only

15. When Type AC or MC cable is installed through holes in metal studs, _____.

     a. a steel plate should be fastened to the stud to protect the cable

     b. a bushing is not required inside the hole in the stud

     c. a bushing is required in the hole if the hole is less than $1\frac{1}{4}$ inches to the face of the stud

     d. the cable must be supported between the studs

16. Disregarding exceptions, where a feeder supplies continuous loads or any combination of continuous and noncontinuous loads, the rating of the overcurrent device shall not be less than the noncontinuous loads plus _____ of the continuous loads.

     a. 80 percent

     b. 100 percent

     c. 125 percent

     d. 150 percent

Notes

Notes

17. Transient voltage surge suppressors (TVSS) may be installed on circuits NOT exceeding _____ volts.

    a.   150

    b.   300

    c.   600

    d. 1000

18. When buried raceways pass under a driveway, the MINIMUM cover requirements _____.

    a.   decrease if installed in rigid metal conduit (RMC)

    b.   do not change in regard to wiring methods used

    c.   shall be increased for direct-buried cables

    d.   can be increased, decreased or remain the same depending on the wiring method used

19. Exposed, nonmetallic surface extensions shall be permitted to run in any direction from an existing outlet, but not within _____ from the floor.

    a.   1 foot

    b.   1½ feet

    c.   2 feet

    d.   2 inches

20. Which one of the following listed luminaires (lighting fixtures) is NOT permitted for installations in clothes closets?

    a.   recessed incandescent

    b.   surface-mounted fluorescent

    c.   pendant luminaire (fixture)

    d.   surface-mounted incandescent

21. Given: an office building contains two-hundred (200), general-purpose, 120 volt, duplex receptacle outlets rated at 15 amperes each. What is the total connected load for the receptacles?

    a. 36,000 VA

    b. 18,000 VA

    c. 54,000 VA

    d. 72,000 VA

22. Which of the following circuit breakers are NOT permitted to be installed in a 120/240 volt, single-phase power panel?

    a. double-pole circuit breakers

    b. bolt-in style circuit breakers

    c. indicating-type circuit breakers

    d. delta-style circuit breakers

23. Regardless of the voltage, the MINIMUM clearance above a diving platform of a swimming pool and service-drop conductors is _____.

    a. 14½ feet

    b. 22½ feet

    c. 19 feet

    d. 21 feet

24. Given: a small retail store has an area of 1,600 square feet. Disregarding exceptions, how many 15 ampere, 120 volt, lighting branch circuits are required for this installation?

    a. 1

    b. 2

    c. 3

    d. 4

Notes

25. An aluminum grounding electrode shall be _____.

   a. installed at least 18 inches above the earth

   b. installed at least 24 inches above the earth

   c. in direct earth burial

   d. prohibited

MASTER ELECTRICIAN
Practice Exam #12

The following questions are based on the 2005 edition of the National Electrical Code® and are typical of questions encountered on most Master Electricians' Exams. Select the best answer from the choices given and review your answers with the answer key included in this book.

ALLOTTED TIME: 75 minutes

**Master Electrician
Practice Exam #12**

1. The MINIMUM size copper grounding electrode conductor tap for a separately derived, alternating-current electrical system with three parallel sets of size 300 kcmil copper secondary conductors shall be _____.

 a. 1/0 AWG

 b. 1 AWG

 c. 2/0 AWG

 d. 3/0 AWG

2. Cables or raceways installed using directional boring equipment shall _____.

 a. be approved for the purpose

 b. have a warning ribbon installed above the cable or raceway

 c. not be permitted

 d. be identified by color coding

3. Given: power for equipment that is directly associated with the radio frequency distribution system is carried by the coaxial cable. The power source is an energy-limiting transformer. The MAXIMUM voltage this coaxial cable may carry is _____ volts.

 a. 50

 b. 60

 c. 120

 d. 150

4. Given: a 15 ampere-rated, general-use, AC snap switch is to serve as a disconnecting means for an AC electric motor. The NEC requires the MAXIMUM full-load current rating of the motor to be no more than _____.

 a. 7.5 amperes

 b. 10 amperes

 c. 12 amperes

 d. 15 amperes

5. In general, the MINIMUM voltage allowed at the line terminals of a 480 volt, three-phase, fire pump controller under motor starting conditions shall be _____.

    a. 466 volts

    b. 456 volts

    c. 432 volts

    d. 408 volts

6. Which of the following statements, if any, is true regarding the illumination for service equipment installed in electrical equipment rooms of commercial or industrial occupancies?

    I.   The illumination shall not be controlled by means of three-way switches.

    II. The illumination shall not be controlled by automatic means only.

    a. I only

    b. II only

    c. both I and II

    d. neither I nor II

7. Given: you have tenant spaces in a retail mall; each occupant shall have access to the main disconnecting means EXCEPT _____.

    a. where the service and maintenance are provided by the building management

    b. where there are more than six disconnecting means provided

    c. where the primary feeder transformer does not exceed 600 volts

    d. where the secondary of the service transformer does not exceed 240 volts to ground

8. Determine the conductor allowable ampacity given the following related information:

   – ambient temperature of 44° C

   – size 250 kcmil THWN copper conductors

   – four (4) current-carrying conductors in the raceway

   – length of raceway is 30 feet

       a. 160 amperes

       b. 167 amperes

       c. 200 amperes

       d. 209 amperes

9. Determine the MAXIMUM overcurrent protection permitted for size 14 AWG, THWN, copper, motor-control circuit conductors tapped from the load side of a motor overcurrent protection device.

   Given: the conductors require short-circuit protection and do not extend beyond the motor-control equipment enclosure.

       a. 20 amperes

       b. 25 amperes

       c. 30 amperes

       d. 100 amperes

10. Which of the following statement(s) is/are TRUE regarding grounding and bonding of metal gas lines?

        a. A buried natural gas line may be used as an electrical systems only grounding conductor.

        b. A buried natural gas line may be used as a grounding electrode when supplemented by another electrode.

        c. The furnace gas line must be bonded to an electrical grounding system.

        d. All of these are true.

11. Given: a grounding ring consists of size 2 AWG bare copper wire encircling a building, buried 36 inches, in direct contact with the earth. The NEC requires the MINIMUM length of the wire for the grounding electrode in this ground ring to be _____.

    a. 15 feet
    b. 20 feet
    c. 25 feet
    d. 50 feet

12. Which one of the following listed single-conductor cable types is NOT permitted by the NEC as a wiring method of solar-photovoltaic source circuits?

    a. SE
    b. UF
    c. USE
    d. SA

13. Which one of the following conductor insulations is identified as oil-resistant?

    a. TW
    c. THE
    b. THWN
    d. MTW

14. The NEC mandates no more than _____ 240 volt, two-pole, branch circuit breakers to be installed in a lighting panelboard.

    a. 12
    b. 20
    c. 21
    d. 42

15. Determine the MINIMUM number of 15 ampere, 120 volt, lighting branch circuits required for a 12,000 square feet multi-family dwelling when each unit has cooking facilities provided.

    a. 15

    b. 20

    c. 24

    d. 30

16. At least one receptacle outlet shall be installed directly above a show window of a retail department store for each _____ linear feet of show window lighting area.

    a. 6

    b. 8

    c. 10

    d. 12

17. When a conduit containing service-entrance conductors runs beneath a building, what is the MINIMUM depth of concrete required to cover a conduit for it to be considered "outside" the building?

    a. 2 inches

    b. 6 inches

    c. 12 inches

    d. 18 inches

18. Continuous duty motors with a marked service factor not less than 1.15 shall have the MINIMUM running overload protection sized at _____ percent of the nameplate rating of the motor.

    a. 115

    b. 125

    c. 130

    d. 140

19. A 90 ampere, 240 volt, single-phase load is located 225 feet from a panelboard and is wired with size 3 AWG, THWN, copper conductors. What is the approximate voltage drop on this circuit?

     a.   6 volts

     b.   4 volts

     c.   8 volts

     d.   10 volts

20. A single-family residence is served with insulated, 120/240 volt, service-drop conductors supported with a grounded messenger wire from the utility company. This service drop crosses the pool; the pool has a diving board that is 5 feet above the water level. The MINIMUM height above water level of the service-drop conductors as permitted by the NEC is _____.

     a.   25 feet

     b.   14½ feet

     c.   22½ feet

     d.   19½ feet

21. At an outdoor motor fuel dispensing facility, the area up to 18 inches above grade and within 20 feet from the edge of an LPG dispenser shall be classified as _____.

     a.   Class I, Division 1

     b.   Class I, Division 2

     c.   Class II, Division 1

     d.   Class II, Division 2

22. In a commercial garage work area, which of the following receptacles, if any, are required to have GFCI protection?

     I.   15 ampere, 120 volt, general-purpose receptacles for hand tools and portable lights

     II. 20 ampere, 120 volt receptacles serving electrical diagnostic equipment only

     a.   I only

     b.   II only

     c.   both I and II

     d.   neither I nor II

23. Given: a feeder is to supply two (2), continuous duty, induction-type, Design B, 208 volt, three-phase, AC motors, one 10 HP and one 7½ HP; the feeder shall have a MINIMUM ampacity of _____.

    a. 55.00 amperes

    b. 68.75 amperes

    c. 62.70 amperes

    d. 38.50 amperes

24. The MINIMUM spacing required between the bottom of a 600 volt rated switchboard and non-insulated busbars mounted in the switchboard cabinet is _____.

    a.   6 inches

    b.   8 inches

    c.   10 inches

    d.   12 inches

25. Given: a rigid metal conduit (RMC) contains only the following three (3) circuits on the load side of the service overcurrent protection devices:

    – two (2) – 150 ampere, three-phase circuits

    – one (1) – 300 ampere, single-phase circuit

    The load side equipment bonding jumper for this conduit must be a MINIMUM size _____ copper.

    a. 1 AWG

    b. 2 AWG

    c. 4 AWG

    d. 6 AWG

MASTER ELECTRICIAN
Practice Exam #13

The following questions are based on the 2005 edition of the National Electrical Code® and are typical of questions encountered on most Master Electricians' Exams. Select the best answer from the choices given and review your answers with the answer key included in this book.

ALLOTTED TIME: 75 minutes

**Master Electrician
Practice Exam #13**

1. Given: a hotel has a stationary electric sign mounted inside a large water fountain. The sign shall be at LEAST _____ feet inside the fountain when measuring from the outside edges of the fountain.

    a. 2
    b. 3
    c. 4
    d. 5

2. Which one of the following statements regarding handhole enclosures is correct?

    a. The enclosure must be sealed to prevent moisture from entering the enclosure.
    b. The enclosure is mounted on a pedestal.
    c. The enclosure must have a hinged cover.
    d. The enclosure must be large enough to permit access by reaching into the enclosure.

3. Branch-circuit conductors supplying outlets for professional-type, arc and Xenon, motion picture projectors shall be at LEAST size _____.

    a. 12 AWG
    b. 10 AWG
    c.   8 AWG
    d.   6 AWG

4. Where rigid nonmetallic conduit (RNC) is installed underground to supply a gasoline dispensing unit, threaded rigid metal conduit (RMC) or threaded intermediate metal conduit (IMC) shall be used for the last _____ feet of the underground run to where the conduit emerges.

    a. 2

    b. 4

    c. 5

    d. 6

Notes

5. Given: a feeder tap conductor is to terminate in a 70 ampere-rated circuit breaker. The ampacity of the feeder tap conductor is required to be NOT less than _____.

    a. 65 amperes

    b. 70 amperes

    c. 85 amperes

    d. 140 amperes

6. When determining the number of conductors permitted to be installed in an outlet box, a reduction of _____ conductors shall be made for one device, two clamps and three grounding conductors.

    a. 3

    b. 4

    c. 5

    d. 6

7. The demand factor of any electrical system is the ratio of the maximum demand for a system to _____.

    a. the total connected load of the system

    b. 125 percent of the total connected load of the system

    c. 125 percent of the total, connected, continuous load of the system

    d. 80 percent of the total, connected, noncontinuous load plus 125 percent of the total, connected, continuous load of the system

Notes

8. All 15 or 20 ampere, single-phase, 125 volt receptacle outlets located within at least _____ of the edge of a fountain shall be provided with GFCI protection.

    a. 10 feet
    b. 15 feet
    c. 20 feet
    d. 25 feet

9. The emergency controls for attended self-service gasoline stations and convenience stores must be located NOT more than _____ feet from the gasoline dispensers.

    a. 20
    b. 50
    c. 75
    d. 100

10. Hospital x-ray equipment supplied by a branch circuit rated at NOT more than _____ amperes may be served by a hard service cord with a suitable attachment plug.

    a. 15
    b. 20
    c. 30
    d. 60

11. What are the MINIMUM size, THWN, copper conductors required to serve a continuous duty, 25 HP, 208 volt, three-phase motor?

    Given: the motor is on the end of a 25 foot conduit run that contains only three (3) conductors at an expected ambient temperature of 50° C.

    a. 6 AWG
    b. 3 AWG
    c. 2 AWG
    d. 1 AWG

12. Which of the following statements regarding the grounding of portable generators is TRUE?

    a. The frame of a portable generator is considered a grounding electrode.

    b. A vehicle-mounted generator must be connected to a driven ground rod.

    c. The equipment grounding terminals of receptacles on a portable generator are isolated from the frame of the generator.

    d. The equipment grounding terminals of receptacles on a portable generator are bonded to the frame of the generator.

13. Which one of the following listed shall be installed on the critical branch of the emergency system in a health care facility?

    a. exit sign

    b. nurse call system

    c. communication system

    d. fire alarm

14. A disconnecting means installed at the distribution point where two or more agricultural buildings are supplied shall NOT be required to _____.

    a. be grounded

    b. contain overcurrent protection

    c. contain a grounded conductor

    d. be rated for the calculated load

15. When calculating the service-entrance conductors for a farm service, the second largest load of the total load shall be computed at not less than _____ percent.

    a. 90

    b. 80

    c. 75

    d. 65

Notes

16. When a building contains multiple elevators, a lighting branch circuit supplying the cars may supply no more than _____car(s).

    a. one

    b. two

    c. three

    d. four

17. What percent of electrically-supplied spaces in a recreational vehicle park must be provided with 30 ampere, 125 volt receptacle outlets?

    a. 60 percent

    b. 70 percent

    c. 90 percent

    d. 100 percent

18. The collector ring used for the grounding of an irrigation machine shall have a current rating of NOT less than _____ of the full-load current of the largest device served plus the sum of the other devices served.

    a. 75 percent

    b. 80 percent

    c. 100 percent

    d. 125 percent

19. Given: a three-phase, 25 kVA rated transformer with a 480 volt primary and a 208Y/120 volt secondary is to be installed. Where primary and secondary protection is provided, what is the MAXIMUM standard ampere rating of secondary overcurrent protection permitted by the NEC?

    a. 80 amperes

    b. 90 amperes

    c. 100 amperes

    d. 110 amperes

20. Where ungrounded conductors are run in parallel in multiple raceways, the equipment grounding conductor, where used, shall be _____.

    a. omitted

    b. run in parallel

    c. installed in one raceway only

    d. bare

21. In agricultural buildings where livestock is housed, any portion of a direct-buried equipment grounding conductor run to the building or structure shall be _____.

    a. insulated or covered copper

    b. bare or insulated copper

    c. bare or insulated aluminum

    d. insulated, covered copper or aluminum

22. Determine the MINIMUM size, THWN, copper, feeder conductors required by the NEC to supply the following 480 volt, continuous duty, three-phase, induction-type, Design B, AC motors.

    – one (1) – 40 HP

    – one (1) – 50 HP

    – one (1) – 60 HP

    a. 2/0 AWG

    b. 3/0 AWG

    c. 4/0 AWG

    d. 250 kcmil

23. Two or more motors may be installed without individual overcurrent protection devices if rated less than 1 HP each and the full-load current rating of each motor does NOT exceed _____ amperes if they are on a 120 volt, single-phase, 20 ampere-rated branch circuit.

    a. 2

    b. 4

    c. 5

    d. 6

Notes

24. Where compressed natural gas vehicles are repaired, the area within ____ inches of the ceiling shall be considered unclassified where adequate ventilation is provided.

    a. 18

    b. 24

    c. 30

    d. 36

25. A three-phase, 480 volt, 100 amp, noncontinuous load is located 390 feet from a panelboard. What MINIMUM size THWN aluminum conductors are required to supply the load if the voltage drop is limited to 3 percent? (K = 21.2)

    a. 2 AWG

    b. 1 AWG

    c. 1/0 AWG

    d. 2/0 AWG

MASTER ELECTRICIAN
Practice Exam #14

The following questions are based on the 2005 edition of the National Electrical Code® and are typical of questions encountered on most Master Electricians' Exams. Select the best answer from the choices given and review your answers with the answer key included in this book.

ALLOTTED TIME: 75 minutes

Master Electrician
Practice Exam #14

1. When a portable generator is used for a portable, optional, standby source and is not considered a separately derived system, the equipment grounding conductor shall be bonded to _____.

   a. the generator frame and the grounded conductor

   b. the system grounding electrode

   c. a grounding electrode only

   d. the grounded conductor

2. Conductors other than service-entrance conductors shall be permitted to be installed in a cable tray with service-entrance conductors provided _____.

   a. a solid fixed barrier of material compatible with the cable tray is installed to separate the systems

   b. the conductors are not in excess of 480 volts to ground

   c. the voltage between the conductors is not in excess of 600 volts

   d. the conductors have equal insulation and temperature ratings

3. When overcurrent protective devices are to be installed to protect fire pumps, they are to be sized to carry the _____ of the fire pump motors.

   a. starting current

   b. full-load running current

   c. stopping current

   d. locked-rotor current

4. How many receptacle outlets are required to be readily accessible in guest rooms and suites of hotels?

   a. 1

   b. 2

   c. 3

   d. 4

5. Which of the following MUST be provided at a patient bed location used for general care in a hospital?

    a. circuit on normal system

    b. circuit on emergency system

    c. "hospital-grade" receptacle outlets

    d. all of these

6. When an electrical service is required to have a grounded conductor present, what is the MINIMUM size grounded conductor permitted for an electrical service using size 1000 kcmil, copper, ungrounded conductors?

    a. 3/0 copper

    b. 2/0 copper

    c. 1/0 copper

    d. 4/0 copper

7. Determine the MINIMUM 20 ampere, 277 volt, lighting branch circuits required for a 150,000 square foot retail department store when the actual connected lighting load is 400 kVA.

    Note: circuit breakers of this size are not rated for continuous use.

    a. 72

    b. 82

    c. 91

    d. 102

8. When a two-gang box contains two (2) single-pole switches (unless the box is equipped with permanently installed barriers), the voltage between the switches shall not be in excess of _____.

    a. 120 volts

    b. 277 volts

    c. 480 volts

    d. 240 volts

Notes

Notes

9. Where portions of a cable, raceway or sleeve are known to be subjected to different temperatures and where condensation is known to be a problem, the raceway or sleeve shall _____.

    a. be filled with an approved material
    b. have an explosionproof seal
    c. be provided with a drain
    d. be installed below the level of the terminations

10. Fluorescent luminaires (light fixtures) installed more than _____ feet above the floor level in patient care areas in hospitals shall NOT be required to be grounded by an insulated equipment grounding conductor.

    a. 6
    b. 6½
    c. 7
    d. 7½

11. When rigid nonmetallic conduit (RNC) is used to enclose conductors supplying a wet-niche luminaire (light fixture) in a swimming pool, a size _____, insulated, copper, grounding conductor shall be installed in the RNC unless a listed, low-voltage lighting system not requiring grounding is used.

    a. 12 AWG
    b. 10 AWG
    c. 8 AWG
    d. 6 AWG

12. The ampacity of phase conductors from the generator terminals to the first overcurrent device shall not be less than _____ percent of the nameplate current rating of the generator where the design of the generator does not prevent overloading.

    a. 100
    b. 115
    c. 125
    d. 150

13. When calculating the total load of an office building and the number of general-purpose receptacle outlets to be installed is not yet determined, an additional load of _____ volt-ampere(s) per square foot is required for the receptacles.

    a. 1

    b. 2

    c. 3

    d. ½

14. An outdoor motor enclosure subject to temporary submersion shall be of a Type _____.

    a. 3RX

    b. 3SX

    c. 4X

    d. 6P

15. Conductors between the controller and the diesel engine of a fire pump are required by the NEC to be _____.

    a. 90° C rated

    b. 104° C rated

    c. stranded

    d. solid

16. The MAXIMUM allowable ampacity of a size 750 kcmil, XHHW, aluminum conductor when there are six (6) current-carrying conductors in the raceway installed in a dry location where the ambient temperature will reach 22° C is _____.

    a. 323.40 amperes

    b. 365.40 amperes

    c. 361.92 amperes

    d. 348.00 amperes

Notes

Notes

17. Cable trays are permitted to be installed in all of the following listed locations EXCEPT _____.

    a. basements

    b. storage rooms

    c. sealed ceiling spaces

    d. when passing through a wall

18. In compliance with the National Electric Code, flat conductor cable (FCC) is permitted to be used for _____.

    I. general-purpose branch-circuit conductors

    II. appliance branch-circuit conductors

    a. I only

    b. II only

    c. both I and II

    d. neither I nor II

19. Determine the MINIMUM size USE aluminum cable permitted for use on an underground, 120/240 volt, single-phase service for a small office building that has a total load of 23,600 VA after all demand factors have been taken into consideration.

    Note: consider all conductor terminations are rated for 75° C.

    a. 1/0 AWG

    b. 2/0 AWG

    c. 1 AWG

    d. 2 AWG

20. The MINIMUM overhead clearance above the water level of a swimming pool for network-powered, broadband communication system conductors shall be _____.

    a. 14 feet

    b. 22½ feet

    c. 25 feet

    d. 27 feet

21. If an apartment complex has a lighting load of 205.4 kVA, what is the demand factor of this lighting load in kVA?

    Given: each apartment unit will contain an electric range (the optional method of calculation is not to be used).

    a. 60.2 kVA

    b. 16.5 kVA

    c. 63.0 kVA

    d. 65.3 kVA

22. All of the following copper conductors are to be installed in an electrical metallic tubing (EMT) that is 40 feet long:

    – Twenty-four (24) – size 10 AWG TW

    – Ten (10) – size 10 AWG THW

    – Fourteen (14) – size 12 AWG THHN

    Determine the MINIMUM trade size of EMT required.

    a. 2 inch

    b. 2½ inch

    c. 3 inch

    d. 3½ inch

23. A type of sign the NEC does NOT require to be listed if installed in conformance with the NEC is a _____ sign.

    a. neon-lighted

    b. portable

    c. HID-lighted

    d. fluorescent-lighted

24. When sizing fuses or circuit breakers for a branch circuit serving a hermatic, refrigerant motor-compressor, the device shall NOT exceed _____ of the rated load current marked on the nameplate of the equipment.

    a. 115 percent

    b. 125 percent

    c. 175 percent

    d. 225 percent

Notes

25. Given: you are installing 90 feet of multi-outlet assembly in the computer lab of a school. The computers are likely to be used simultaneously. Determine the MINIMUM number of 20 ampere, 120 volt, single-phase branch circuits required to supply the multi-outlet assembly.

    a. 4

    b. 5

    c. 6

    d. 7

MASTER ELECTRICIAN
Practice Exam #15

The following questions are based on the 2005 edition of the National Electrical Code® and are typical of questions encountered on most Master Electricians' Exams. Select the best answer from the choices given and review your answers with the answer key included in this book.

ALLOTTED TIME: 75 minutes

Master Electrician
Practice Exam #15

1. The time-delay feature required for a legally required, standby generator set, to avoid retransfer in case of short-time re-establishment of the normal power source, shall permit a setting of _____.

    a. one minute

    b. five minutes

    c. ten minutes

    d. fifteen minutes

2. In general, when conductors of different insulation ratings are installed in a common raceway and the voltage is 600 volts or less, the NEC requires _____.

    a. all conductors shall have an insulation rating equal to at least the maximum circuit voltage applied to any conductor in the raceway

    b. the conduit allowable fill to be limited to 31 percent

    c. the conductors of separate ratings to be grouped and bound

    d. the insulation of the lower-rated conductors to be identified with blue or yellow colors

3. An emergency system is required to have _____ seconds to have power available in the event of failure of the normal supply.

    a. 10

    b. 15

    c. 45

    d. 60

4. Given: a direct burial cable has a circuit voltage of 45 kV. The NEC mandates the MINIMUM burial depth of the cable to be _____.

    a. 24 inches

    b. 36 inches

    c. 42 inches

    d. 48 inches

5. Elevator driving motors used with a generator field control are rated as _____ duty motors.

    a. intermittent

    b. continuous

    c. variable

    d. controlled

6. When installed in a metal enclosure, the grounding conductor of a transient voltage surge suppressor (TVSS) shall be _____.

    a. a minimum size of 12 AWG copper

    b. a minimum size of 10 AWG copper

    c. bonded to one end of the enclosure

    d. bonded to both ends of the enclosure

7. Given: an RV campground has a total of one hundred and fifty (150) campsites with electrical power. Twenty-five (25) of the campsites are reserved as tent sites. How many of the sites are required to be furnished with a 20 ampere, 120 volt receptacle outlet?

    a. 105

    b. 125

    c. 150

    d. None

Notes

Notes

8. Determine the MINIMUM size type SO cord permitted to supply a 40 HP, 460 volt, three-phase, continuous duty, AC, wound rotor motor installed in an area with an ambient temperature of 86° F.

    a. 2 AWG

    b. 4 AWG

    c. 6 AWG

    d. 8 AWG

9. Determine the full-load current of a 30 HP, three-phase, 230 volt, synchronous motor with a 90 percent power factor.

    a. 63.0 amperes

    b. 69.3 amperes

    c. 76.23 amperes

    d. 86.62 amperes

10. A single-family dwelling unit has three ovens rated at 6, 8 and 3.5 kW, a cooktop rated at 6 kW and a broiler rated at 3.5 kW. The demand factor, in kW, to be calculated for this kitchen equipment on the ungrounded service-entrance conductors is _____.

    a. 12.2 kW

    b. 18.6 kW

    c. 27.3 kW

    d. 30.1 kW

11. When installing emergency, battery pack, lighting unit equipment (emergency lights), the branch circuit supplying this equipment shall _____.

    a. be connected to the nearest receptacle outlet

    b. come from the nearest outlet of power that is compatible with the emergency lights rated voltage

    c. be fed only from an identified emergency lighting panel

    d. be on the same branch circuit serving the normal lighting in the area

12. In an industrial establishment, what is the MAXIMUM length of a 200 ampere-rated busway that may be tapped to a 600 ampere-rated busway without providing additional overcurrent protection?

    a. 10 feet

    b. 25 feet

    c. 50 feet

    d. 75 feet

13. If a fuse label is not marked with a fuse interrupting rating, then the fuse size _____.

    a. has an interrupting rating of 100,000 amperes and is current limiting

    b. has an interrupting rating of 100,000 amperes and is not current limiting

    c. has an interrupting rating of 10,000 amperes

    d. has no interrupting rating

14. Given: a recreational vehicle park has a total of 40 RV sites supplied with electrical power. How many sites are required to be equipped with at least one (1), 50 ampere, 125/250 volt receptacle outlet?

    a. 40

    b. 20

    c. 8

    d. 2

15. When wireways longer than 5 feet are mounted horizontally, they shall be secured on each end or joint and shall be secured at intervals not exceeding _____.

    a. 3 feet

    b. 4 feet

    c. 5 feet

    d. 10 feet

Notes

16. The MINIMUM burial depth required for conduit or cables installed under an airport runway, concourse or tarmac is _____.

    a. 1½ feet

    b. 2 feet

    c. 3 feet

    d. 4 feet

17. Determine the MINIMUM size overload protection required for a 480 volt, three-phase, 15 HP, continuous duty, AC motor given the following conditions:

    – Design B

    – temperature rise is 40° C

    – service factor is 1.12

    – actual nameplate rating is 18 amperes

    a. 20.7 amperes

    b. 18.0 amperes

    c. 22.5 amperes

    d. 23.4 amperes

18. Given: the overload protection you have selected on the above motor is not sufficient to start the motor and trips. Determine the MAXIMUM size overload protection permitted.

    a. 22.5 amperes

    b. 20.7 amperes

    c. 25.2 amperes

    d. 23.4 amperes

19. Given: a 50 kVA transformer has a 480 volt, three-phase primary and a 208/120 volt, three-phase secondary. Overcurrent protection is required on both the primary and secondary side of the transformer. Determine the MAXIMUM, standard size, overcurrent protection device permitted on the primary side.

    a. 125 amperes
    b. 150 amperes
    c. 175 amperes
    d. 200 amperes

20. What is the MINIMUM demand load, in VA, for the general-purpose receptacles in an office building with a total of one hundred-fifty (150), 15 and 20 ampere, 120 volt receptacle outlets?

    a. 18,500 VA
    b. 10,000 VA
    c. 27,000 VA
    d. 13,500 VA

21. Determine the MAXIMUM number of size 12 AWG conductors that may be installed in a 3½ inch deep, three-gang, masonry box that contains three switches.

    a. 21
    b. 23
    c. 24
    d. 27

22. A horizontal raceway entering a dust-ignitionproof enclosure from one that is not does not require a seal-off if it is _____ in length.

    a. 18 inches
    b. 5 feet
    c. 10 feet
    d. 25 inches

Notes

23. A transformer is to be installed in a retail mall to serve a tenant space. A grounding electrode conductor for the separately derived system is required to be connected to the nearest available, effectively grounded _____ in the building.

    a. underground metal gas pipe

    b. structural metal member

    c. service equipment enclosure

    d. metal raceway

24. In a dairy, non-current-carrying metal parts of equipment MUST be grounded by _____.

    a. a copper equipment grounding conductor

    b. a driven ground rod

    c. a metal water pipe

    d. a grounded conductor

25. A dry-type transformer of less than 600 volts and _____ is permitted to be installed in a hollow space of a building, such as above a lift-out ceiling, provided there is adequate ventilation.

    a.   25 kVA

    b.   37½ kVA

    c.   50 kVA

    d.   112½ kVA

SIGN ELECTRICIAN
Practice Exam #16

The following questions are based on the 2005 edition of the National Electrical Code® and are typical of questions encountered on most Journeyman and Master Sign Electricians' Exams. Select the best answer from the choices given and review your answers with the answer key included in this book.

ALLOTTED TIME: 75 minutes

Sign Electrician
Practice Exam #16

1. Which of the following conductor insulation types listed is NOT suitable for use in a rigid nonmetallic conduit (RNC) buried underground in direct contact with the earth?

    a. TW

    b. THHN

    c. THWN

    d. XHHW

2. As long as it has continuity to the grounding electrode system, an equipment grounding conductor may be _____.

    a. rigid metal conduit (RMC)

    b. a structural member of the building steel frame

    c. the building metallic water system

    d. a driven ground rod

3. The NEC mandates at LEAST _____ of free conductor to be left at each junction box for the purpose of splicing conductors or making connections to luminaires (lighting fixtures) or devices.

    a. 4 inches

    b. 6 inches

    c. 8 inches

    d. 10 inches

4. Given: a 20 ampere, 120 volt, single-phase branch circuit is supplying an electric sign. What is the MAXIMUM voltage drop the NEC recommends on this branch circuit?

    a. 2.4 volts

    b. 3.6 volts

    c. 5.0 volts

    d. 6.0 volts

5. Which of the following, if any, may help to reduce voltage drop in a branch circuit?

    I.   Install the conductors in a larger raceway.

    II. Increase the size of the ungrounded conductors.

        a. I only

        b. II only

        c. either I or II

        d. neither I nor II

6. When installing neon secondary circuit conductors over 1000 volts, metal parts of a building shall be permitted as _____.

        a. a secondary return conductor

        b. an equipment grounding conductor

        c. an ungrounded conductor

        d. none of these

7. Given: you are to install a rigid metal conduit (RMC) buried underground in direct contact with the earth under a parking lot of a retail mall to supply an electric sign. What is the MINIMUM burial depth of the conduit?

        a.   6 inches

        b. 12 inches

        c. 18 inches

        d. 24 inches

8. Portable or mobile electric signs shall be provided with factory installed _____.

        a. GFCI protection

        b. AFCI protection

        c. FGCI protection

        d. GACI protection

9. The required operating current for a specific luminous tube is dependent on which of the following?

    a. junction box length

    b. diameter of tubing

    c. distance from adjacent tubing

    d. branch circuit size

10. The MINIMUM size conductor permitted for wiring neon secondary circuits rated at 1000 volts or less is _____.

    a. 14 AWG

    b. 12 AWG

    c. 16 AWG

    d. 18 AWG

11. The MAXIMUM number of size 8 AWG conductors with THW insulation permitted to be installed in a trade size ¾ inch rigid nonmetallic conduit (RNC) more than 2 feet in length is _____.

    a. 3

    b. 4

    c. 5

    d. 6

12. A portable electric sign shall not be placed within a pool or fountain or within _____ from the inside walls of the pool or fountain.

    a. 5 feet

    b. 10 feet

    c. 15 feet

    d. 20 feet

13. Fixed or stationary electric signs installed within a fountain at a hotel are required to be at LEAST ____ from the outside edges of the fountain.

     a. 10 feet

     b. 8 feet

     c. 6 feet

     d. 5 feet

14. Given: an electric sign is to be installed in the parking lot of a retail store and is not provided with barriers to protect it from physical damage; the NEC requires the sign to be at LEAST ____ above the areas accessible to vehicular traffic.

     a. 10 feet

     b. 12 feet

     c. 14 feet

     d. 16 feet

15. Branch circuits that supply electric signs or outline lighting systems shall be rated not to exceed ____.

     a. 15 amperes

     b. 20 amperes

     c. 30 amperes

     d. 50 amperes

16. Of the following listed, which one of the connector types is prohibited for connection of an equipment grounding conductor to electric signs or their disconnecting means?

     a. sheet metal screw

     b. pressure connector

     c. clamp

     d. lug

17. Given: you are to install a service for a highway billboard electric sign consisting of three (3), two-wire branch circuits. The service disconnecting means shall have a rating of at LEAST _____.

   a. 15 amperes

   b. 30 amperes

   c. 60 amperes

   d. 100 amperes

18. Branch-circuit conductors installed in ballast compartments of fluorescent luminaires (lighting fixtures) installed in electric signs that are within 3 inches of a ballast shall have a temperature rating of NOT less than _____.

   a. 105° C

   b. 90° C

   c. 75° C

   d. 60° C

19. Where branch-circuit conductors are installed in raceways, the conductors larger than size _____ are required to be stranded conductors.

   a. 10 AWG

   b.   8 AWG

   c. 12 AWG

   d. 14 AWG

20. Switches or similar devices controlling a transformer in an electric sign must have an ampere rating of not less than _____ percent of the ampere rating of the transformer.

   a. 100

   b. 200

   c. 125

   d. 300

21. An electric sign load is considered to be continuous if the maximum current of the sign is expected to continue for _____ hour(s) or more.

    a. ½
    b. 1
    c. 2
    d. 3

22. When installing a sign that contains a 208 volt luminaire (lighting fixture) outside of an office building window, the NEC requires the sign be not less than _____ feet from the window.

    a. 3
    b. 6
    c. 8
    d. 10

23. The MAXIMUM allowable ampacity of size 8 AWG, THWN, copper conductors installed in a 50 foot length of rigid nonmetallic conduit (RNC) is _____ when there are not more than three (3) current-carrying conductors in the conduit.

    a. 40 amperes
    b. 50 amperes
    c. 55 amperes
    d. 60 amperes

24. Each commercial building accessible to pedestrians shall have an outside electric sign branch circuit rated at LEAST _____ amperes that supplies no other load.

    a. 15
    b. 20
    c. 25
    d. 30

Notes

25. What is the MAXIMUM distance permitted by the NEC that a required disconnect may be located from an electric sign?

    a. 10 feet

    b. 12 feet

    c. 25 feet

    d. 50 feet

SIGN ELECTRICIAN
Practice Exam #17

The following questions are based on the 2005 edition of the National Electrical Code® and are typical of questions encountered on most Journeyman and Master Sign Electricians' Exams. Select the best answer from the choices given and review your answers with the answer key included in this book.

ALLOTTED TIME: 75 minutes

Sign Electrician
Practice Exam #17

1. Given: you have installed a transformer in the attic of a department store and the transformer serves an electric sign installed outdoors in the front of the building. This installation is in compliance with the NEC if there is an access door of at LEAST _____ in size.

    a. 2 feet × 2 feet

    b. 3 feet × 2 feet

    c. 3 feet × 3 feet

    d. 4 feet × 3 feet

2. When determining the proper size branch-circuit conductors required to supply an electric sign installed with fluorescent luminaires, the calculated load shall be based on _____.

    a. the total watts of the lamps

    b. 1 VA per linear inch of the lamps

    c. the total ampere rating of the luminaires

    d. 10 VA per linear foot of the lamps

3. The NEC requires _____ receptacle outlet(s) to be installed on an electric sign.

    a. no

    b. one

    c. two

    d. three

4. A type of electric sign the NEC does NOT require to be listed, if installed in conformance with the Code, is a _____ sign.

    a. neon-lighted

    b. portable

    c. highway

    d. fluorescent-lighted

5. When an electric sign is installed within a fountain or within at least _____ of the fountain edge, all branch circuits supplying the sign shall have GFCI protection.

    a.   6 feet

    b. 10 feet

    c. 20 feet

    d. 50 feet

6. The maximum ampere rating of an overcurrent protection device, such as a circuit breaker or a fuse protecting a branch circuit, is directly dependent upon _____.

    a. voltage

    b. impedance

    c. current flow

    d. switch size

7. Determine the MINIMUM size listed metal box permitted by the NEC that contains two (2), size 12 AWG conductors and three (3), size 8 AWG conductors.

    a.   1 inch $\times$ 1¼ inch round

    b.   3 inch $\times$ 2 inch $\times$ 1½ inch device

    c.   4 inch $\times$ 1½ inch octagon

    d.   3 inch $\times$ 2 inch $\times$ 2½ inch device

8. What is the MAXIMUM length allowed for cords supplying portable electric signs in dry locations?

    a. 10 feet

    b.   6 feet

    c. 15 feet

    d. 25 feet

Notes

9. The overcurrent protection for size 10 AWG copper conductors, regardless of the temperature rating of the insulation, shall NOT exceed _____ amperes when protecting nonmotorized electric sign loads.

    a. 15
    b. 20
    c. 25
    d. 30

10. Bonding conductors provided for electric signs shall be sized at not less than _____ copper conductors.

    a. 14 AWG
    b. 12 AWG
    c. 10 AWG
    d.   8 AWG

11. Given: you have a roof-mounted, metal sign enclosure made of sheet steel. This metal sign enclosure is required to be a MINIMUM thickness of _____.

    a. .202 inch
    b. .020 inch
    c. .160 inch
    d. .016 inch

12. In general, flexible metal conduit (FMC) shall be supported at least every _____ feet and within _____ inches of an outlet box, panelboard or disconnect switch.

    a. 4½, 12
    b. 6, 12
    c. 8, 18
    d. 10, 18

13. Rigid nonmetallic conduit (RNC) shall be securely fastened within
    _____ foot/feet of an outlet box, junction box or panelboard.

    a. 1

    b. 2

    c. 2½

    d. 3

14. If there are no junction boxes provided, what is the MAXIMUM
    number of 90° bends permitted in a run of rigid nonmetallic conduit
    (RNC) used to enclose branch-circuit conductors supplying an
    electric sign?

    a. two

    b. three

    c. four

    d. six

15. Given: a commercial building accessible to pedestrians has a
    branch circuit supplying an electric sign that contains neon tubing
    installations only. This branch circuit shall be rated NO more than
    _____ amperes.

    a. 15

    b. 20

    c. 30

    d. 40

16. The NEC requires lamps for outdoor lighting to be located below
    all energized conductors or other utilization equipment. An
    exception to this requirement would be which of the following?

    a. The lamps shall be within 6½ feet of the ground level.

    b. A lockable disconnecting means must be provided.

    c. Conductors are to be identified by orange insulation.

    d. The lamps must have an isolated grounding conductor.

Notes

17. Disregarding exceptions, when sign enclosures are supported by metal poles and the poles contain supply conductors to the sign, a handhole not less than _____ in size is required at the base of the pole to provide access to the supply terminations.

    a. 2 inch × 4 inch
    b. 4 inch × 4 inch
    c. 4 inch × 6 inch
    d. 6 inch × 6 inch

18. Which of the following is required, if any, for open branch-circuit conductors that extend through the exterior wall of a building?

    I. The conductors are required to be in a metallic sleeve.
    II. Sleeves through the wall must slant upward from the outside of the building to the inside of the building.

    a. I only
    b. II only
    c. both I and II
    d. neither I nor II

19. What expected change in length, due to expansion, is a 200 foot run of rigid nonmetallic conduit (RNC) to have when the RNC is installed outside and is exposed to an annual 90° temperature variation from the warmest day to the coldest day?

    a. 3.65 inches
    b. 7.30 inches
    c. 10.96 inches
    d. 3.75 inches

20. The MAXIMUM number of ungrounded conductors allowed to be terminated in an individual terminal of a circuit breaker installed in a panelboard is _____.

    a. one conductor
    b. two conductors
    c. three conductors
    d. four conductors

21. Determine the MINIMUM trade size intermediate metal conduit (IMC) permitted by the NEC to enclose three (3), size 8 AWG, copper, branch-circuit conductors in a 95 foot conduit run.

   a. ½ inch

   b. ¾ inch

   c. 1 inch

   d. 1¼ inch

22. An electrical time clock provided for lighting branch circuits is usually connected in _____ with a single lighting circuit to be controlled.

   a. series

   b. parallel

   c. sequence

   d. tandem

23. An electrical device used to reduce voltage without changing the available power is a/an _____.

   a. rectifier

   b. amplifier

   c. transformer

   d. capacitor

24. What is the MINIMUM size copper conductor required for an overhead conductor span when supplying a pole-mounted electric sign having a total load of 1,800 VA?

   Given:

   – 120 volt branch circuit

   – 25 feet between supports

   – no messenger wire is provided

   a. 8 AWG

   b. 10 AWG

   c. 12 AWG

   d. 14 AWG

25. Given: a pole-mounted electric sign is to be installed at a service station with gasoline dispensing units. The sign is to be supplied by branch-circuit conductors installed in rigid nonmetallic conduit (RNC). In order for the RNC to be outside of the area classified as a hazardous location, the conduit is required to be at LEAST _____ from the gasoline dispensing units.

    a. 20 feet

    b. 25 feet

    c. 30 feet

    d. 50 feet

RESIDENTIAL ELECTRICIAN
Final Exam

The following questions are based on the 2005 edition of the National Electrical Code® and are typical of questions encountered on most Residential Electricians' Exams. Select the best answer from the choices given and review your answers with the answer key included in this book. Passing score on this exam is 70 percent. The exam consists of 60 questions valued at 1.67 points each, so you must answer at least 42 questions correct for a passing score. If you do not score at least 70 percent, keep studying and try again. GOOD LUCK.

ALLOTTED TIME: 3 hours

Residential Electrician
Final Exam

1. The NEC mandates specific branch circuits, receptacle outlets and/or utilization equipment to be provided with a ground-fault circuit interrupter (GFCI); this device is intended _____.

    a. to prevent overloading the branch-circuit conductors

    b. for the protection of equipment from ground faults

    c. for the protection of personnel

    d. to prevent overloading the circuit breaker

2. Where nonmetallic-sheathed cable is installed in nonmetallic device boxes, the cable sheath shall extend not less than _____ inside the box and beyond any cable clamp.

    a. ⅜ inch

    b. ¾ inch

    c. ½ inch

    d. ¼ inch

3. When replacing a luminaire (light fixture) in an outlet that does not provide an equipment grounding conductor, which of the following is/are permitted?

    a. Install a listed luminaire and add GFCI protection to the branch circuit.

    b. Insulate the equipment grounding conductor of the luminaire.

    c. Connect the equipment grounding conductor of the luminaire to the ungrounded conductor of the branch circuit.

    d. Connect the equipment grounding conductor of the luminaire to the grounded conductor of the branch circuit.

4. Given: a 240 volt, 17 kW rated electric range is to be installed in the kitchen of a single-family dwelling unit. Determine the MINIMUM size copper NM cable the NEC requires to supply this electric range.

    a. 8 AWG

    b. 6 AWG

    c. 4 AWG

    d. 2 AWG

5. When sizing the overcurrent protection required for nonmotor-operated appliances such as electric ranges, water heaters and cook-tops, the overcurrent protection shall NOT exceed _____ percent of the rated current of the appliance. Where this value does not correspond to a standard overcurrent device ampere rating, the next higher standard rating shall be permitted.

    a. 115

    b. 125

    c. 150

    d. 175

6. When nails are used to fasten metal boxes to wooden studs and the nails pass through the interior of the box, the nails shall be within _____ inch of the back of the box.

    a. ¼

    b. ½

    c. ⅜

    d. ¾

7. The largest size copper conductor permitted for nonmetallic-sheathed cable is _____.

    a. 4 AWG

    b. 1 AWG

    c. 2/0 AWG

    d. 2 AWG

Notes

Notes

8. For the purpose of determining conductor fill in a device box, the NEC mandates a switch to be counted as equal to two conductors. The volume allowance for the two conductors shall be based on _____.

    a. the largest wire in the box

    b. the largest grounding conductor in the box

    c. the largest wire connected to the switch

    d. the number of clamps in the box

9. When a single-family dwelling is supplied with a single-phase, three-wire, 120/240 volt service drop from the local utility company, the service disconnecting means shall have a rating of at LEAST _____.

    a.   60 amperes

    b. 100 amperes

    c. 150 amperes

    d. 200 amperes

10. The ampacity of type NM cable shall be in accordance with the _____ conductor temperature rating; the _____ temperature rating shall be permitted to be used for ampacity derating purposes.

    a. 60° C, 90° C

    b. 75° C, 90° C

    c. 60° C, 75° C

    d. 90° F , 75° C

11. For residential electric ranges having a rating of 8.75 kW or more, the branch circuit rating shall be at LEAST _____.

    a. 30 amperes

    b. 40 amperes

    c. 50 amperes

    d. 60 amperes

12. In a dwelling unit where no wall adjacent to a bathroom sink is available to install a receptacle outlet, the outlet is permitted to be on the front of the vanity below the countertop provided it is no more than _____ from the outside edge of the sink.

    a. 12 inches

    b. 18 inches

    c. 24 inches

    d. 36 inches

13. Switching devices must be located at least _____ feet from the inside walls of a swimming pool unless separated from the pool by a solid fence, wall or other permanent barrier.

    a. 5

    b. 10

    c. 15

    d. 20

14. In dwelling units, 15 or 20 ampere-rated, 125 volt receptacle outlets installed within six feet of the outside edge of a sink in a laundry room shall _____.

    a. have a weather-proof cover

    b. have AFCI protection

    c. be a single receptacle

    d. be provided with GFCI protection

15. Range hoods shall be permitted to be cord-and-plug-connected if the receptacle outlet is accessible and supplied by an individual branch circuit and the length of the cord does NOT exceed _____.

    a. 36 inches

    b. 24 inches

    c. 18 inches

    d. 12 inches

Notes

16. Single-family dwellings with direct outdoor grade level access in front and back are required to have _____.

   I. one receptacle outlet at the back

   II. one receptacle outlet at the front

   a. I only

   b. II only

   c. both I and II

   d. neither I nor II

17. Underground service conductors that are not encased in concrete and buried 18 inches or more below grade shall have their location identified by a warning ribbon that is placed in the trench at least _____ above the underground installation.

   a. 6 inches

   b. 12 inches

   c. 18 inches

   d. 24 inches

18. When the white-colored, insulated conductor in an NM cable is used for a three-way switch loop, the conductor shall be _____.

   a. permanently re-identified

   b. a return conductor

   c. used only to supply switched outlets

   d. a grounding conductor only

19. When a 20 ampere, 120 volt receptacle outlet provided for laundry appliances is installed in the bathroom of a dwelling unit, the receptacle is required to be _____.

   a. not readily accessible

   b. a single receptacle

   c. GFCI protected

   d. fed from the bathroom basin outlets

20. When a single branch circuit with an equipment grounding conductor is supplying a detached garage from a house, what, if any, additional grounding electrode(s) is/are required at the garage?

    a. a driven ground rod and underground metal water pipe

    b. a metal underground water pipe only

    c. a metal underground water pipe and building steel, if available

    d. No additional electrodes are required.

21. Which of the following, if any, is permitted for closing an unused circuit breaker opening in panelboards, load centers and switchboards?

    I. Install an unused or spare circuit breaker that is compatible with the equipment.

    II. Close the opening by installing a closure plate that fills the opening and fits securely.

    a. I only

    b. II only

    c. either I or II

    d. neither I nor II

22. In dwellings, lighting outlets shall be permitted to be controlled by occupancy sensors provided they are _____.

    a. automatic

    b. located in the hallway

    c. within 6 feet of the door(s)

    d. equipped with a manual override

23. Which one of the following rooms in a dwelling is NOT required to have a wall switch-controlled lighting outlet?

    a. bathroom

    b. bedroom

    c. kitchen

    d. attached garage

24. Where nonmetallic-sheathed cable is used with nonmetallic boxes no larger than 2¼ inches × 4 inches, the cable is not required to be secured to the box if the cable is fastened within at LEAST ____ of the box.

    a. 6 inches

    b. 8 inches

    c. 12 inches

    d. 10 inches

25. A receptacle that provides power to a pool recirculating pump motor shall be permitted not less than ____ from the inside wall of the pool.

    a. 5 feet

    b. 4 feet

    c. 8 feet

    d. 10 feet

26. Receptacle outlets installed on a temporary service pole at a construction site are required to be ____.

    a. of the nongrounding type

    b. of the grounding type and be provided with GFCI protection

    c. rated 20 amperes

    d. of the twist-lock type and rated at least 20 amperes

27. Given: a direct-buried, landscape lighting circuit to be installed carries 24 volts and type UF cable is to be used. The cable must be buried with an earth cover of at least ____.

    a. 6 inches

    b. 12 inches

    c. 18 inches

    d. 24 inches

28. For dwelling units, the 125 volt, 15 and 20 ampere-rated receptacle outlets for kitchen countertop surfaces are to be supplied with at LEAST _____ small-appliance branch circuit(s) rated at _____.

    a. one (1), 20 amperes

    b. two (2), 20 amperes

    c. three (3), 15 amperes

    d. three (3), 20 amperes

29. When trade size ⅜ inch flexible metal conduit (FMC) is used as a fixture whip from an outlet box to a luminaire (light fixture), the flex shall not exceed _____ in length.

    a. 10 feet

    b. 4 feet

    c. 5 feet

    d. 6 feet

30. A cord-and-attachment-plug-connected room air conditioner shall not exceed _____ of the rating of the branch circuit where no other loads such as lighting fixtures are supplied.

    a. 80 percent

    b. 75 percent

    c. 60 percent

    d. 50 percent

31. Track lighting or ceiling fans are not to be located within a MINIMUM of _____ vertically from the top of the bathtub rim or shower stall threshold.

    a. 6 feet

    b. 8 feet

    c. 4 feet

    d. 10 feet

Notes

32. When NM cable is installed through slots or holes in metal framing members, the NEC requires which of the following for the protection of the cable?

    a. Listed bushings or grommets installed after cable installation.

    b. Listed bushings or grommets installed before cable installation.

    c. Split bushings or grommets installed after cable installation.

    d. The NEC permits any of these installations.

33. What is the MAXIMUM length of a flexible cord that may be used for a recirculating pump motor installed on a swimming pool at a multi-family dwelling?

    a.   3 feet

    b.   4 feet

    c.   6 feet

    d.   10 feet

34. Holiday lighting installations for dwellings are permitted to be in use for no more than _____.

    a.   30 days

    b.   60 days

    c.   90 days

    d.   120 days

35. Where a switch is installed in an outlet box and is wired with 12/2 NM cable and no equipment grounding conductor exists and the switch has a metal faceplate, which of the following must be done in order for this installation to be in compliance with the NEC?

    I.   Replace the metal faceplate with a nonmetallic faceplate.

    II. Provide the branch circuit with GFCI protection.

    a.   I only

    b.   II only

    c.   either I or II

    d.   both I and II

36. When driving a ground rod at a 45° angle and rock bottom is encountered, the electrode shall be permitted to be buried in a trench that is at LEAST _____ deep.

    a. 36 inches

    b. 30 inches

    c. 48 inches

    d. 24 inches

37. What is the MAXIMUM distance a 125 volt receptacle outlet may be installed from a hot tub installed outside at a dwelling?

    a. 20 feet

    b. 15 feet

    c. 10 feet

    d. 5 feet

38. Panelboards are NOT permitted to be installed in which of the following areas of dwellings?

    a. wet locations

    b. bathrooms

    c. bedrooms

    d. detached garages

39. What is the MINIMUM number of 20 ampere-rated branch circuits required for a single-family dwelling?

    a. 4

    b. 2

    c. 5

    d. 3

Notes

40. Using the standard method of calculation for a single-family dwelling, determine the MINIMUM demand load, in VA, on the ungrounded service-entrance conductors when the residence has the following fixed appliances installed:

    – water heater – 4,800 VA

    – dishwasher – 1,200 VA

    – garbage disposal – 1,150 VA

    – trash compactor – 800 VA

    – attic fan – 1,200 VA

    a.   6,863 VA

    b.   9,150 VA

    c.   8,579 VA

    d.   11,438 VA

41. In general, for size 10 AWG, copper, nonmetallic-sheathed cable (NM), the overcurrent protection shall not exceed _____.

    a.   25 amperes

    b.   20 amperes

    c.   35 amperes

    d.   30 amperes

42. Given: a single-family dwelling having a computed demand load of 175 amperes is provided with a 120/240 volt, single-phase, three-wire service drop from the local utility company. What is the MINIMUM size aluminum conductors with THWN insulation permitted for use as ungrounded service-entrance conductors?

    a.   1/0 AWG

    b.   2/0 AWG

    c.   3/0 AWG

    d.   4/0 AWG

43. When a 20 ampere-rated branch circuit in a residence supplies only fixed, resistance-type baseboard heaters, this circuit may be loaded to a MAXIMUM value of _____.

    a. 16 amperes

    b. 20 amperes

    c. 18 amperes

    d. 14 amperes

44. If an evaporative cooler or an air conditioner is mounted on the roof of a multi-family dwelling unit, where is the service receptacle outlet for the unit to be located?

    a. within 75 feet of the unit

    b. within 50 feet of the unit and on the same level

    c. within 25 feet of the unit and on the same level

    d. The outlet is not required for dwelling units.

45. What is the MINIMUM size copper equipment grounding conductor required for a 15 ampere-rated branch circuit installed in rigid metal conduit (RMC) that supplies a swimming pool circulating pump motor?

    a. 14 AWG

    b. 12 AWG

    c. 10 AWG

    d. 8 AWG

46. For a one-family dwelling, at least one receptacle outlet shall be installed in each detached garage if the garage is _____.

    a. more than 750 square feet in size

    b. less than 20 feet from the dwelling

    c. more than 20 feet from the dwelling

    d. provided with electric power

Notes

47. The service grounding electrode conductor is sized by the rating of _____.

     a. the main circuit breaker

     b. the service-drop conductors

     c. the service-entrance conductors

     d. the ground rod

48. When exceptions are not taken into consideration, duplex receptacles installed in a residential garage must have _____.

     a. a weatherproof cover

     b. a double-pole circuit breaker

     c. a dedicated circuit

     d. GFCI protection

49. When wall-mounted, the required 125 volt, 15 and 20 ampere-rated, general-purpose receptacle outlets in dwelling units shall not be located more than _____ above the floor.

     a. 5½ feet

     b. 4 feet

     c. 18 inches

     d. 4½ feet

50. When a dwelling unit has a hallway 25 feet in length, the NEC requires at LEAST _____ receptacle outlet(s) to be installed in the hallway.

     a. 1

     b. 2

     c. 3

     d. 4

51. Given: a two-gang device box contains two (2), size 12/2 with ground NM cables connected to a duplex receptacle and two (2), size 14/2 w/ground NM cables connected to a single-pole switch. The two-gang box also contains four clamps. The device box is required to have a volume of at LEAST _____.

    a. 28 cubic inches

    b. 30 cubic inches

    c. 34 cubic inches

    d. 36 cubic inches

52. What is the allowable ampacity of a size 14/3 AWG, Type SJO flexible cord when only two of the conductors are considered current-carrying?

    a. 15 amperes

    b. 18 amperes

    c. 20 amperes

    d. 25 amperes

53. When Type NM cable is run on unfinished basement walls to switches or receptacles, the cable is required to be protected from physical damage by means of _____.

    I. electrical metallic tubing (EMT)

    II. rigid nonmetallic conduit (RNC)

    a. I only

    b. II only

    c. either I or II

    d. neither I nor II

54. Where UF cable is used as a substitute for Type NM cable for interior wiring, the conductor insulation is required to be rated at _____.

    a. 60° C

    b. 75° C

    c. 90° C

    d. 100° C

Notes

55. Nonmetallic-sheathed cable shall be permitted to be installed unsupported where the cable is _____.

    I. fished in existing walls of finished buildings
    II. not more than 4½ feet from the last cable support to the connection point of a luminaire (lighting fixture)

       a. I only
       b. II only
       c. neither I nor II
       d. either I or II

56. Given: a size 8 AWG NM cable is used to supply a 5,000 watt, 240 volt, residential cook-top. Determine the MAXIMUM standard size circuit breaker permitted for overcurrent protection for this appliance.

       a. 30 amperes
       b. 35 amperes
       c. 20 amperes
       d. 40 amperes

57. Given: you are to install several 1,000 watt, 240 volt, 6 feet long baseboard heaters in a residence. Determine the MAXIMUM number of heaters that may be supplied by a single 20 ampere, 240 volt, single-phase branch circuit.

       a. 1
       b. 2
       c. 3
       d. 4

58. All wet-niche luminaries (lighting fixtures) installed in swimming pools shall be removable from the water for _____.

    I. relamping or inspection while on the pool deck

    II. maintenance or servicing from a dry location

        a. I only

        b. II only

        c. neither I nor II

        d. either I or II

59. For maintenance, the disconnecting means for associated swimming pool _____ shall be readily accessible and within sight from the equipment to be disconnected.

    I. pump motors

    II. luminaires

        a. I only

        b. II only

        c. both I and II

        d. neither I nor II

60. When a device box is installed in a wall having a gypsum (drywall) surface, those surfaces that are broken or incomplete around the device box employing a flush-type faceplate shall be repaired so there are no gaps or open spaces greater than _____ at the edge of the box.

        a. ⅛ inch

        b. ¼ inch

        c. ⅜ inch

        d. ½ inch

Notes

Notes

JOURNEYMAN ELECTRICIAN
Final Exam

The following questions are based on the 2005 edition of the National Electrical Code® and are typical of questions encountered on most Journeyman Electricians' Exams. Select the best answer from the choices given and review your answers with the answer key included in this book. Passing score on this exam is 70 percent. The exam consists of 75 questions valued at 1.33 points each, so you must answer at least 53 questions correct for a passing score. If you do not score at least 70 percent, keep studying and try again. GOOD LUCK.

ALLOTTED TIME: 4 hours

Journeyman Electrician
Final Exam

1. When determining the initial, standard size, time-delay fuse for branch-circuit, short-circuit and ground-fault protection for a three-phase, continuous duty, squirrel cage, AC motor, the fuse shall have a rating of _____ of the FLC of the motor. When the value determined does not correspond to a standard ampere rating, the next higher standard rating shall be permitted.

   a.   175 percent

   b.   150 percent

   c.   225 percent

   d.   250 percent

2. In compliance with the NEC, a conductor with three white stripes on a black background is permitted to be used as a/an _____ conductor.

   a.   ungrounded

   b.   grounding

   c.   grounded

   d.   hot-phase

3. An installation consisting of seven (7), non-jacketed, size 12 AWG, Type MC cables, each with three (3) current-carrying conductors bundled together longer than 24 inches, shall have an adjustment factor of _____.

   a.   45 percent

   b.   50 percent

   c.   60 percent

   d.   70 percent

4. Which one of the following listed wiring methods is NOT permitted to be installed in ducts or plenums fabricated to transport environmental air?

    a. flexible metallic tubing

    b. type MI cable

    c. electrical metallic tubing

    d. liquidtight flexible metal conduit

5. When wiring gasoline dispensing pumps, the first fitting that should be installed in the raceway that emerges from below ground or concrete into the base of the gasoline dispenser is a/an _____.

    a. automatic cut-off breakaway valve

    b. disconnect

    c. sealing fitting

    d. no fittings of any kind are permitted in this location

6. One kVA is equal to _____.

    a. one hundred watts

    b. one thousand volt-amperes

    c. one hundred kilowatts

    d. one hundred volt-amperes

7. A 15 kW, 208 volt, single-phase heat pump will have a full-load current rating of _____.

    a. 42 amperes

    b. 125 amperes

    c. 65 amperes

    d. 72 amperes

Notes

8. Heavy-duty lighting track is lighting track identified for use exceeding _____.

    a.　15 amperes

    b.　600 watts

    c.　20 amperes

    d.　120 volts

9. In general, threaded entries into explosionproof equipment shall be made up with at least _____ threads fully engaged.

    a.　4

    b.　5

    c.　5½

    d.　6

10. Live parts of generators operated at more than _____ volts to ground shall not be exposed to accidental contact where accessible to unqualified persons.

    a.　24

    b.　50

    c.　120

    d.　150

11. Unless listed and identified otherwise, terminations and conductors of circuits rated 100 amperes or larger shall be rated at _____.

    a.　60° C

    b.　75° C

    c.　90° C

    d.　120° C

12. A lighting and appliance branch-circuit panelboard is one having a MINIMUM of more than _____ percent of its overcurrent devices protecting lighting and appliance branch circuits.

    a. 10

    b. 20

    c. 30

    d. 40

13. Regardless of the voltage of the service-entrance conductors, a clearance of not less than _____ shall be maintained between the maximum water level of a swimming pool and the service-entrance conductors.

    a. 10 feet

    b. 14½ feet

    c. 19 feet

    d. 22½ feet

14. Given: a galvanized eye-bolt is to be used as the point of attachment for an electrical service. The NEC requires the eye-bolt to be installed not less than _____ feet above finished grade.

    a. 8

    b. 12

    c. 10

    d. 15

15. Under which one of the following conditions are you required to use an approved, electrically-conductive, corrosion-resistant compound on metal conduit threads?

    a. in Class I, Division 1 locations

    b. in Class II, Division 1 locations

    c. on field cut threads for indoor locations

    d. on field cut threads in corrosive locations

Notes

16. Disregarding exceptions, each patient bed location in critical care areas of hospitals shall be supplied by _____.

    I. one or more branch circuits from the emergency system

    II. one or more branch circuits from the normal system

        a. I only

        b. II only

        c. either I or II

        d. both I and II

17. Freestanding-type, cord-and-plug-connected office partitions shall not contain more than _____, 15 ampere, 125 volt receptacle outlets.

        a. 6

        b. 13

        c. 10

        d. 8

18. Which one of the following listed raceways is NOT permitted for use of enclosing control wiring for electric, motor-driven fire pumps?

        a. flexible metal conduit (FMC)

        b. liquidtight flexible metal conduit (LFMC)

        c. liquidtight flexible nonmetallic conduit (LFNC)

        d. intermediate metal conduit (IMC)

19. What MINIMUM size bonding conductor must bond all metal parts associated with a hydro-massage bathtub, including all metal piping systems, metal parts of the electrical equipment and pump motors?

        a. 12 AWG

        b. 10 AWG

        c. 8 AWG

        d. 6 AWG

20. When installing galvanized rigid metal conduit where structural members readily permit fastening, the conduit is required to be fastened within _____ of a junction box or panelboard.

    a. 3 feet
    b. 5 feet
    c. 6 feet
    d. 10 feet

21. In general, flexible cords are prohibited from being installed in raceways; an exception to this rule is when ____.

    a. supplying luminaires (light fixtures) in a school
    b. installed in an elevator shaft of an office building
    c. installed in flexible metal conduit (FMC) above a lift-out ceiling
    d. used for protection from physical damage for industrial installations

22. Given: a 240 volt, wye-connected motor derives power from a converted single-phase source. It is desired to connect a 120/240 volt transformer for control of the motor on the load side of the converter. The control transformer must _____.

    a. be connected to the manufactured phase
    b. be disconnected after start-up of the motor
    c. have separate overcurrent protection
    d. not be connected to the manufactured phase

23. What is the MAXIMUM size overcurrent protection device required to protect size 14 AWG conductors used for the pump motor-control circuit that is protected by the motor branch-circuit protection device and extends beyond the enclosure?

    a. 15 amperes
    b. 20 amperes
    c. 45 amperes
    d. 100 amperes

Notes

Notes

24. Branch circuits supplying more than one motor shall have an ampacity of at least _____ percent of the FLC of the largest motor and 100 percent of the FLC of the other motor(s) in the group.

    a. 100
    b. 115
    c. 125
    d. 150

25. For ladder-type cable trays where all the cables are sized 1000 kcmil or larger, the cables _____.

    a. must be installed in a single layer
    b. are permitted to be installed in multiple layers if not bound together
    c. shall be installed in multiple layers
    d. shall have an outer shield

26. Given: you are to install fifty (50), general-purpose receptacle outlets in an existing office having a 120/240 volt, single-phase electrical system. The receptacles are to be rated 20 amperes each, 120 volts. Determine the current in amperes the receptacles will add to each of the ungrounded service-entrance conductors.

    a. 37.50 amperes
    b. 75.00 amperes
    c. 125 amperes
    d. 150 amperes

27. Refer to the previous question and determine the MINIMUM number of 20 ampere, 120 volt branch circuits required for this installation.

    a. 16
    b. 12
    c. 8
    d. 4

28. When twenty (20), size 12 AWG, copper, current-carrying conductors with THHN insulation are installed in a 75 foot run of trade size 1½ inch electrical metallic tubing (EMT), what is the allowable ampacity rating of each conductor?

    a. 15 amperes

    b. 20 amperes

    c. 25 amperes

    d. 30 amperes

29. Non-current-carrying metal parts of electrical equipment shall be kept at least _____ away from lightning rod conductors, or they shall be bonded to the lightning rod conductors if not separated by dense materials such as concrete, brick or wood.

    a. 3 feet

    b. 4 feet

    c. 5 feet

    d. 6 feet

30. The MINIMUM size copper conductor that may be used for grounding the secondary of an instrument transformer is _____.

    a. 6 AWG

    b. 8 AWG

    c. 10 AWG

    d. 12 AWG

31. Surface metal raceways and surface nonmetallic raceways that have removable covers are permitted to contain splices and taps; the fill is NOT to exceed _____ of the area of the raceway at the point of the splice or tap.

    a. 54 percent

    b. 60 percent

    c. 75 percent

    d. 50 percent

Notes

32. Control-circuit conductors and motor branch-circuit conductors are permitted by the NEC to occupy the same raceway if the control-circuit conductors _____.

    a. have an applied voltage not exceeding 24 volts

    b. have a voltage to ground not to exceed 150 volts

    c. have a voltage to ground not to exceed 277 volts

    d. are functionally associated with the motor system

33. Which one of the following statements regarding handhole enclosures is correct?

    a. The enclosure must be designed as to prevent moisture from entering.

    b. The enclosure is mounted above ground on a concrete pedestal.

    c. The enclosure must be constructed of concrete.

    d. The enclosure must be large enough to permit access by reaching into the enclosure.

34. A feeder supplying two (2), continuous duty, 208 volt, three-phase, AC motors, one rated 10 HP and one rated 7½ HP, shall have a MINIMUM ampacity of _____.

    a. 55.00 amperes

    b. 68.75 amperes

    c. 62.70 amperes

    d. 38.50 amperes

35. The reason the NEC requires all phase conductors of the same circuit to be in the same metal raceway is to reduce _____.

    a. expense

    b. inductive heat

    c. voltage drop

    d. resistance

36. Which of the following conductor insulation types, if any, is/are acceptable for wiring in a fluorescent luminaire (lighting fixture) when the branch-circuit conductor passes within three inches of the ballast?

    I.   THW

    II. THWN

        a.   I only

        b.   II only

        c.   both I and II

        d.   neither I nor II

37. Where only Class 1 circuit conductors are in a raceway, the derating factors given in 310.15(B)(2)(a) of the NEC shall apply only if the conductors _____.

        a.   have an insulation rating of 60° C

        b.   do not have an insulation rating of 90° C

        c.   are of a size 14 AWG or less

        d.   carry continuous loads in excess of 10 percent of the ampacity of each conductor

38. A general-purpose receptacle outlet is permitted to be installed within a bathtub or shower space _____.

        a.   if GFCI protected

        b.   if AFCI protected

        c.   if provided with a watertight cover

        d.   never

39. An attachment plug and receptacle may be permitted to serve as a motor controller if the motor is portable and has a MAXIMUM rating of _____.

        a.   ⅛ HP

        b.   ¼ HP

        c.   ⅓ HP

        d.   ½ HP

Notes

Notes

40. Given: you are installing a 200 ampere-rated temporary service at a construction site and the only grounding electrode available is a driven ground rod. The MINIMUM size grounding electrode conductor to ground the service to the ground rod is _____ copper.

   a. 8 AWG

   b. 4 AWG

   c. 6 AWG

   d. 3 AWG

41. Surface metal raceways are required to be secured and supported at intervals NOT exceeding _____.

   a.   4 feet

   b.   8 feet

   c.   10 feet

   d.   the manufacturer's instructions

42. All spa and hot tub equipment installed in non-dwelling occupancies shall be provided with an emergency shutoff or control switch within sight of the unit(s) and located not less than _____ from the inside wall of the spa or hot tub.

   a.   6 feet

   b.   10 feet

   c.   5 feet

   d.   12 feet

43. Feeders to floating buildings are permitted to be installed in _____ where flexible connections are required.

   a. MC cable

   b. AC cable

   c. NMC cable

   d. portable power cable

44. Given: a recreational vehicle park has fifty (50) RV sites with electrical power. How many sites are required to be equipped with at least one (1), 30 ampere, 125 volt receptacle outlet?

   a. 10

   b. 15

   c. 35

   d. 50

45. In a marina, the disconnecting means for a boat receptacle outlet must be readily accessible and not more than _____ from the receptacle outlet.

   a. 6 feet

   b. 10 feet

   c. 30 inches

   d. 50 feet

46. When installing exposed rigid metal conduit (RMC) vertically from industrial machinery, if the conduit is made up with threaded couplings, the distance between supports of the conduit shall NOT exceed _____.

   a. 3 feet

   b. 5 feet

   c. 10 feet

   d. 20 feet

47. Cablebus shall be permitted to be installed only for _____ locations.

   a. exposed

   b. commercial

   c. concealed

   d. hazardous

Notes

48. Outlet boxes intended to enclose flush devices shall have an internal depth of _____.

    a. $\frac{1}{2}$ inch

    b. $\frac{15}{16}$ inch

    c. $\frac{7}{8}$ inch

    d. $1\frac{1}{2}$ inches

49. In general, with respect to the service disconnecting means, the grounding electrode conductor shall be connected to the grounded service conductor on _____.

    I. the supply side

    II. the load side

    a. I only

    b. II only

    c. either I or II

    d. both I and II

50. Given: a single grounding electrode is used for a commercial building. The measured resistance is greater than 25 ohms. What, if anything, is required?

    a. Add additional electrodes until the resistance is less than 25 ohms.

    b. Add one additional electrode.

    c. Nothing is required by the NEC.

    d. Pour salt around the electrode.

51. Fire pump supply conductors on the load side of the final disconnecting means and overcurrent device(s) are permitted to be routed through a building if _____.

    a. encased in at least 2 inches of brick or concrete

    b. installed in raceways with power conductors

    c. concealed within walls, floors or ceilings that provide a thermal barrier of material that has at least a 15-minute finish rating

    d. installed in a listed electrical circuit protective system with a 1-hour fire rating

52. All 125 volt, single-phase, 15 and 20 ampere-rated receptacle outlets installed outdoors in a public space that is accessible to the public must _____.

    I. be GFCI protected

    II. have an enclosure that is weatherproof

    a. I only
    b. II only
    c. both I and II
    d. neither I nor II

Notes

53. Nonconductive optical fiber cable may occupy the same raceway with conductors supplying lights and power provided the operating voltage does NOT exceed _____.

    a. 480 volts
    b. 600 volts
    c. 250 volts
    d. 120 volts

54. How many receptacle outlets are required to be installed directly above a show window of a department store having a show window width of 36 feet?

    a. 2
    b. 3
    c. 4
    d. 5

55. When branch-circuit conductors are connecting one or more units of a data-processing system to a source of supply, the branch circuits shall have an ampacity of at LEAST _____ of the total connected load.

    a. 80 percent
    b. 115 percent
    c. 125 percent
    d. 150 percent

Notes

56. The metal frame of a building or structure can serve as a grounding electrode under which, if any, of the following conditions?

    I. Ten feet or more of a single structural metal member is in direct contact with the earth or encased in concrete that is in direct contact with the earth.

    II. The structural metal is bonded to two driven ground rods if the ground resistance of a single ground rod exceeds 25 ohms.

    a. I only

    b. II only

    c. either I or II

    d. neither I nor II

57. Control-circuit devices with screw-type pressure terminals used with size 14 AWG or smaller copper conductors shall be torqued to at LEAST _____ pound-inches, unless identified for a different torque value.

    a. 9

    b. 7

    c. 6

    d. 5

58. Straight runs of trade size 2 inch rigid metal conduit (RMC) that is made up with threaded couplings shall be supported at LEAST every _____.

    a. 10 feet

    b. 14 feet

    c. 16 feet

    d. 20 feet

59. The allowable ampacity of a size 4 AWG, THW, copper conductor installed in a conduit with three (3) other current-carrying conductors of the same size and insulation in a location with an ambient temperature of 98° F is _____.

    a. 60 amperes

    b. 75 amperes

    c. 80 amperes

    d. 85 amperes

60. In general, electrical nonmetallic tubing (ENT) shall be securely fastened in place at LEAST every _____.

    a. 3 feet

    b. 5 feet

    c. 4 feet

    d. 8 feet

61. Given: a metal electrical junction box with a flat metal blank plate contains the following conductors:

    – six (6) – size 10 AWG with THWN insulation

    – ten (10) – size 12 AWG with THWN insulation

    – three (3) – size 14 AWG equipment grounding conductors

    Determine the MINIMUM size junction box, in cubic inches, that is required by the NEC.

    a. 36 cubic inches

    b. 49 cubic inches

    c. 43 cubic inches

    d. 40 cubic inches

62. Where equipment grounding is required when installing electrical nonmetallic tubing (ENT), a separate equipment grounding conductor _____.

I.   shall be installed in the raceway

II. shall be permitted to be installed outside the raceway

    a.   I only

    b.   II only

    c.   either I or II

    d.   neither I nor II

63. Enclosures containing circuit breakers and motor controllers installed in a Class II, Division 2 hazardous location shall be _____.

    a.   dust-tight

    b.   heavy-duty type

    c.   rain-tight

    d.   general-duty type

64. Given: a 20 ampere-rated branch circuit in a residence supplies only fixed, resistance-type, baseboard electric heaters. This branch circuit may be loaded to a MAXIMUM value of _____.

    a.   14 amperes

    b.   16 amperes

    c.   18 amperes

    d.   20 amperes

65. When an inverse-time circuit breaker is installed only as a disconnecting means of a single continuous duty motor, the circuit breaker must have an ampere rating of a LEAST _____ of the full-load running current of the motor.

    a.   100 percent

    b.   125 percent

    c.   115 percent

    d.   150 percent

66. The allowable ampacity of a size 750 kcmil, XHHW, aluminum conductor, when there are six (6) current-carrying conductors in the raceway having a length of more than 24 inches installed in a dry location where the ambient temperature is 22° C, is _____.

    a. 323.40 amperes

    b. 365.40 amperes

    c. 361.92 amperes

    d. 348.00 amperes

67. When calculating the service-entrance conductors for a farm service, the second largest load of the total load shall be computed at _____.

    a. 90 percent

    b. 80 percent

    c. 65 percent

    d. 75 percent

68. Direct-buried conductors and cables emerging from the ground and up a pole shall be protected to a point above finished grade of at LEAST _____.

    a. 6 feet

    b. 8 feet

    c. 10 feet

    d. 12 feet

69. Junction boxes used in straight pulls of conductors of sizes 4 AWG and larger shall be sized so the length of the box shall not be less than _____ times the trade diameter of the largest raceway entering the box.

    a. 4

    b. 6

    c. 8

    d. 10

Notes

Notes

70. After the furniture is installed in guest rooms or guest suites in hotels or motels, how many receptacle outlets are required to be readily accessible?

   a. 1

   b. 2

   c. 3

   d. 4

71. Where track lighting is installed in a continuous row, each individual section of not more than ___ feet in length shall be securely supported.

   a. 2

   b. 4

   c. 6

   d. 8

72. When conductors are adjusted in size to compensate for voltage drop, equipment grounding conductors, where required, shall be adjusted proportionally according to _____.

   a. circular mil area

   b. diameter in inches

   c. temperature rating

   d. area in square inches

73. A multi-wire branch-circuit supply to a gasoline dispensing pump shall be provided with a switch that will disconnect _____.

   a. only one ungrounded supply conductor

   b. all of the ungrounded supply conductors only

   c. only the neutral (grounded) conductor

   d. the grounded conductor and all of the ungrounded supply conductors

74. Which one of the following listed wiring methods is NOT approved for use in an agricultural establishment?

    a. NMC cable

    b. UF cable

    c. NM-B cable

    d. SE cable

75. On a temporary service when a rigid metal conduit (RMC) is used as a driven grounding electrode, it shall _____.

    a. be supplemented by an additional grounding electrode if the resistance to ground is less than 25 ohms

    b. be a minimum trade size of ¾ inch

    c. have an outer coating of corrosion-preventing material

    d. be prohibited

Notes

Notes

MASTER ELECTRICIAN
Final Exam

The following questions are based on the 2005 edition of the National Electrical Code® and are typical of questions encountered on most Master Electricians' Exams. Select the best answer from the choices given and review your answers with the answer key included in this book. Passing score on this exam is 75 percent. The exam consists of 75 questions valued at 1.33 points each, so you must answer at least 57 questions correct for a passing score. If you do not score at least 75 percent, keep studying and try again. GOOD LUCK.

ALLOTTED TIME: 4 hours

Notes

Master Electrician
Final Exam

1. Given: after all demand factors have been taken into consideration for an office building, the demand load is determined to be 90,000 VA. The building has a 120/240 volt, single-phase electrical system. If the service-entrance conductors are to be installed in a buried rigid nonmetallic conduit (RNC), what MINIMUM size copper conductors with THHW insulation are required for the ungrounded service-lateral conductors?

    a. 400 kcmil

    b. 350 kcmil

    c. 300 kcmil

    d. 500 kcmil

2. Given: a commercial building is to be supplied from a transformer having a 480Y/277 volt, three-phase primary and a 208Y/120 volt, three-phase secondary. The secondary will have a balanced computed demand load of 416 amperes per phase. The transformer is required to have a MINIMUM kVA rating of _____.

    a. 100 kVA

    b. 150 kVA

    c. 86 kVA

    d. 200 kVA

3. When a motor controller enclosure is installed outdoors and is subject to be exposed to sleet, it shall have a MINIMUM rating of type _____.

    a. 3

    b. 3S

    c. 3R

    d. 3SX

Notes

4. Which of the following installations would not require a disconnect within sight of the motor if the motor and motor controller are NOT within sight of each other?

    a. The disconnect is capable of being locked in the OPEN position.

    b. The disconnecting means is impractical or introduces additional hazards.

    c. Industrial installations equipped with written safety procedures and maintained by qualified personnel.

    d. All of these apply.

5. Which of the following statement(s) is/are true, if any, regarding fire pump motors installed at a hotel?

    I. A fire pump motor shall be permitted to be supplied by a separate service.

    II. A fire pump motor shall have automatic protection against overloads.

    a. I only

    b. II only

    c. both I and II

    d. neither I nor II

6. Health care facilities require ground-fault protection _____.

    a. on all 208Y/120 volt, three-phase, four-wire systems

    b. on all essential electrical system feeders

    c. to be omitted from all life-safety branch circuits

    d. on the next level of feeders after the service disconnect

7. What MINIMUM voltage is required after one and one-half (1½) hours to the emergency lighting supplied from a storage battery if the normal source voltage of 120 volts is discontinued?

    a. 60 volts

    b. 90 volts

    c. 105 volts

    d. 120 volts

Notes

8. Determine the MAXIMUM ampere rating for an overload protective device responsive to motor current, as permitted by the NEC, used to protect a 20 HP, 230 volt, three-phase, induction type, AC motor with a temperature rise of 48° C and a full-load ampere rating of 54 amperes indicated on the nameplate.

    a. 54.0 amperes

    b. 70.2 amperes

    c. 62.1 amperes

    d. 75.6 amperes

9. Hazardous (classified) atmospheres containing combustible metal dust such as aluminum or magnesium are considered to be in material Group _____ classifications.

    a. C

    b. D

    c. E

    d. F

10. Given: a 15 foot horizontal run of rigid metal conduit (RMC) is to be installed between two enclosures located in a Class II, Division 1 hazardous location. One enclosure is dust-ignitionproof, the other is not. The NEC requires at LEAST _____ sealing fitting(s).

    a. none required

    b. one

    c. two

    d. three

11. Given: a junction box contains the following:

    – three (3) – size 6 AWG, one grounded and two ungrounded conductors

    – one (1) – size 8 AWG equipment grounding conductor

    – two (2) – internal clamps

    – one (1) – pigtail

    Determine the MINIMUM size junction box, in cubic inches, the NEC requires.

    a.   16 cubic inches

    b.   18 cubic inches

    c.   20 cubic inches

    d.   23 cubic inches

12. Determine the MINIMUM required ampacity of the conductors supplying an elevator motor given the following related information:

    – 5 HP, 208 volts, three-phase, wound rotor

    – 15 minute rated

    – nameplate current rating is 18 amperes

    a.   15.3 amperes

    b.   16.2 amperes

    c.   18.0 amperes

    d.   22.5 amperes

13. Heating elements of cables shall be separated from the edge of outlet boxes that are to be used for mounting surface luminaires (lighting fixtures) by a MINIMUM distance of _____.

    a.   12 inches

    b.   10 inches

    c.   8 inches

    d.   6 inches

Notes

14. Nonmetallic underground conduit with conductors (NUCC) larger than trade size _____ shall not be used.

    a. 2 inch

    b. 3 inch

    c. 4 inch

    d. 6 inch

15. Ceiling-suspended luminaires (lighting fixtures) or paddle fans located _____ or more above the maximum water level of an indoor installed spa or hot tub shall not require GFCI protection.

    a. 10 feet

    b. 7½ feet

    c. 8 feet

    d. 12 feet

16. When flat conductor cable (FCC) is used for general-purpose branch circuits, the MAXIMUM rating of the circuits shall be _____ amperes.

    a. 20

    b. 30

    c. 15

    d. 10

17. Lamps located in cellulose nitrate storage vaults shall be installed in rigid luminaires (lighting fixtures) of the _____ type.

    a. glass-enclosed and gasketed

    b. vaporproof

    c. explosionproof

    d. polyvinyl-enclosed and gasketed

18. Infrared heating lamps rated at _____ watts or less shall be permitted with lampholders of the medium-base, porcelain type.

     a. 300

     b. 350

     c. 400

     d. 500

19. General-purpose receptacle outlets installed in the bathroom of a hotel or motel must be _____.

     I. GFCI protected

     II. on a separate 20 ampere-rated branch circuit

     a. I only

     b. II only

     c. neither I nor II

     d. both I and II

20. What type of luminaire (light fixture) must be used in a totally-enclosed poultry house when condensation may be present?

     a. explosionproof

     b. watertight

     c. suitable for damp locations

     d. nonmetallic

21. When sizing overcurrent protection for fire pump motors, the device(s) shall be selected or set to carry the _____ of the motor.

     a. locked-rotor current

     b. full-load running current

     c. full-load amperage as indicated on the motor nameplate

     d. starting current

22. Relay conductors having an applied voltage of 24 volts and associated with a motor shall be _____.

    a. permitted to occupy the same raceway with the motor

    b. run in a separate raceway

    c. blue in color

    d. not smaller than size 12 AWG

23. When calculating the total load for a mobile home park before demand factors are taken into consideration, each individual mobile home lot shall be calculated at a MINIMUM of _____.

    a. 20,000 VA

    b. 15,000 VA

    c. 24,000 VA

    d. 16,000 VA

24. A main circuit breaker mounted inside a switchboard or panelboard must have its center of the grip of the operating handle, when in the highest position, not more than _____ above the floor.

    a. 6 feet

    b. 6 feet 6 inches

    c. 6 feet 7 inches

    d. 7 feet

25. Which of the following classified (hazardous) location(s), if any, does the NEC permit flexible metal conduit (FMC) for connections to motors?

    I. Class I, Division 2

    II. Class II, Division 1

    a. I only

    b. II only

    c. both I and II

    d. neither I nor II

26. A type of wiring method NOT acceptable for use within agricultural buildings is _____.

    a. rigid nonmetallic conduit (RNC)

    b. NMC cable

    c. jacketed type MC cable

    d. open wiring on insulators

27. Nonmetallic surface extensions shall be permitted to be run in any direction from an existing outlet, but not within _____ of the floor level.

    a. 1 foot

    b. 1½ feet

    c. 2 feet

    d. 2 inches

28. Which of the following electrical equipment, if any, is permitted to be installed in cellulose nitrate film storage vaults?

    a. portable lights

    b. receptacle outlets

    c. light switches

    d. none of these

29. Where liquidtight flexible metal conduit (LFMC) is installed in a Class I, Division 2 hazardous location, it shall be permitted to delete the bonding jumper if the LFMC is not longer than _____ feet and the overcurrent protection in the circuit is limited to _____ amperes or less and the load is not a power utilization load.

    a. 6, 20

    b. 10, 10

    c. 6, 10

    d. 6, 15

Notes

Notes

30. Which one of the following wiring methods is NOT permitted to be installed in ducts or plenums used for environmental air?

    a. rigid metal conduit (RMC)

    b. electrical metallic tubing (EMT)

    c. liquidtight flexible metal conduit (LFMC)

    d. flexible metal conduit (FMC)

31. Luminaires (lighting fixtures) shall be supported independently of an outlet box where the weight exceeds _____ pounds unless the outlet box is listed for the weight to be supported.

    a. 50

    b. 40

    c. 30

    d. 20

32. Color coding shall be permitted to identify intrinsically safe conductors where they are colored _____ and where no other conductors of the same color are used.

    a. light blue

    b. red

    c. yellow

    d. purple

33. The NEC permits a building to have more than one service when _____.

    I. the load requirements of the building are at least in excess of 800 amperes

    II. the building is separated by firewalls with a four hour rating

    a. I only

    b. II only

    c. either I or II

    d. neither I nor II

34. Given: the kitchen of a restaurant contains the following listed cooking-related equipment:

    – one (1) – 14 kW range

    – one (1) – 5.0 kW water heater

    – one (1) – 0.75 kW mixer

    – one (1) – 2.5 kW dishwasher

    – one (1) – 2.0 kW booster heater

    – one (1) – 2.0 kW broiler

    Determine the demand load, in kW, on the service-entrance conductors for the kitchen equipment after applying the demand factors.

    a. 19.00 kW

    b. 26.25 kW

    c. 18.38 kW

    d. 17.06 kW

35. If a retail store to be built will have a total of one hundred (100), single-phase, 277 volt, lighting branch circuits, the store is required to have at LEAST _____ lighting panelboard(s) to supply the 277 volt lighting loads.

    a. 1

    b. 2

    c. 3

    d. 4

36. What is the MINIMUM size equipment grounding conductor required for a 5 HP, three-phase, 208 volt, continuous duty motor having 20 ampere rating overload protection and short-circuit overcurrent protection rated at 30 amperes?

    a. 10 AWG

    b. 14 AWG

    c. 12 AWG

    d. 8 AWG

37. The NEC requires ventilation of a battery room where batteries are being charged to prevent _____.

    a. battery corrosion

    b. electrostatic charge

    c. deterioration of the building steel

    d. an accumulation of an explosive mixture

38. What is the MAXIMUM balanced demand load, in VA, permitted to be connected to a new service of a commercial building given the following conditions?

    – The service is 208Y/120 volts, three-phase, four-wire with a 600 ampere-rated main circuit breaker.

    – The maximum load must not exceed 80 percent of the ampere rating of the main circuit breaker.

    a. 57,600 VA

    b. 99,840 VA

    c. 172,923 VA

    d. 178,692 VA

39. Determine the MAXIMUM standard size overcurrent protection required for the primary and secondary side of a transformer when primary and secondary overcurrent protection is to be provided given the following related information:

    – 150 kVA rating

    – Primary – 480 volt, three-phase, three-wire

    – Secondary – 208Y/120 volt, three-phase, four-wire

    a. Primary – 500 amperes, Secondary – 500 amperes

    b. Primary – 450 amperes, Secondary – 600 amperes

    c. Primary – 500 amperes, Secondary – 450 amperes

    d. Primary – 450 amperes, Secondary – 500 amperes

40. Openings around electrical penetrations of a wall of a computer room are required to be _____.

    a. insulated

    b. airtight

    c. fire-stopped

    d. soundproof

41. The local utility company is to supply a single-family dwelling with a 120/240 volt, single-phase, messenger-supported service drop that extends directly above the swimming pool. If the pool has a diving board 2 feet above the water level, the service-drop conductors must have a clearance of at LEAST _____ above the diving board.

    a. 14½ feet

    b. 16½ feet

    c. 20½ feet

    d. 22½ feet

42. In a retail shopping mall, all 125 volt, single-phase, 15 and 20 ampere-rated receptacle outlets located _____ are required to be provided with GFCI protection.

    I. in bathrooms

    II. outdoors and accessible to the public

    a. I only

    b. II only

    c. neither I nor II

    d. both I and II

43. Conductors installed in conduit exposed to direct sunlight in close proximity to rooftops have been shown, under certain conditions, to experience an increase in temperature of _____ above ambient temperature.

    a. 30° F

    b. 30° C

    c. 60° F

    d. 60° C

Notes

Notes

44. Given: a 6,000 square feet bank building is to be constructed and the exact number of general-purpose receptacle outlets to be located in the building is unknown; the bank will also require a sign circuit. Determine the MINIMUM number of 120 volt, 20 ampere-rated branch circuits required by the NEC for the general lighting, receptacle outlets and the sign.

    a. 12
    b. 13
    c. 14
    d. 15

45. Enclosures containing circuit breakers, switches and motor controllers located in Class II, Division 2 hazardous locations shall be listed as ____.

    a. gastight
    b. vaporproof
    c. dusttight
    d. stainless steel

46. Data-processing equipment is permitted to be connected to a branch circuit by flexible cord and attachment plug cap if the cord does NOT exceed _____ in length.

    a. 6 feet
    b. 8 feet
    c. 10 feet
    d. 15 feet

47. Where required, conduit seals installed in Class I, Division 1 and 2 locations shall have the minimum thickness of the sealing compound not less than the trade size of the sealing fitting and in no case less than _____.

    a. ½ inch
    b. ⅝ inch
    c. ¾ inch
    d. 1 inch

48. Given: a single-family dwelling to be built will have 4,000 square foot of livable space, a 600 square foot garage, a 400 square foot porch, a 2,000 square foot unfinished basement (adaptable for future use), three (3) small-appliance branch circuits and one (1) branch circuit for the laundry room. Determine the demand load, in VA, on the ungrounded service-entrance conductors for the general lighting and receptacle loads using the standard method of calculation for a single-family dwelling.

    a. 10,350 VA

    b. 9,825 VA

    c. 7,350 VA

    d. 24,000 VA

49. Determine the MINIMUM size 75° C rated conductors permitted to supply a 25 HP, 208 volt, three-phase, Design C, fire pump motor.

    a. 4 AWG

    b. 3 AWG

    c. 2 AWG

    d. 1 AWG

50. Refer to the previous question. Determine the MAXIMUM standard size overcurrent protection device the NEC permits to protect the fire pump motor.

    a. 200 amperes

    b. 250 amperes

    c. 350 amperes

    d. 450 amperes

51. Determine the initial, permitted, operational setting of an adjustable, inverse time circuit breaker used for branch-circuit, short-circuit and ground-fault protection of a 10 HP, 208 volt, three-phase, squirrel-cage, Design B, continuous duty motor.

    a. 30.8 amperes

    b. 35.7 amperes

    c. 77.0 amperes

    d. 338.8 amperes

52. Each motor controller shall be capable of _____ and _____ the motor it controls.

    a. starting, stopping

    b. starting, running

    c. stopping, disconnecting

    d. stopping, interrupting

53. When viewing the busbars from the front of a three-phase switchboard or panelboard, the "A" phase _____.

    a. is permitted to be any busbar

    b. shall be the busbar on the left

    c. shall be the busbar on the right

    d. shall be the center busbar

54. The MINIMUM size copper grounding electrode conductor for an AC service supplied with four parallel sets of size 500 kcmil aluminum conductors is _____.

    a. 2/0 AWG

    b. 3/0 AWG

    c. 4/0 AWG

    d. 250 kcmil

55. The NEC permits a listed fluorescent luminaire (light fixture) to be cord-and-plug-connected provided the cord is _____.

    a. readily accessible

    b. limited to 3 feet in length

    c. visible its entire length

    d. GFCI protected

56. Which of the following wiring methods, if any, is permitted for use in a duct of a kitchen hood in a restaurant?

    a. MI cable

    b. rigid metal conduit (RMC)

    c. electrical metallic tubing (EMT)

    d. none of these

57. Given: you are to build a 200 ampere, 120/240 volt, single-phase, three-wire, overhead service for a commercial occupancy in an area where the ambient temperature reaches 110° F. Determine the MINIMUM size copper conductors required for the service-entrance conductors. Consider all terminations and conductors to be rated at 75° C.

    a. 250 kcmil

    b. 4/0 AWG

    c. 3/0 AWG

    d. 2/0 AWG

58. For a commercial or industrial establishment, what is the MINIMUM headroom required in front of a main service disconnect switch?

    a. 6 feet

    b. 6 feet 6 inches

    c. 8 feet

    d. 10 feet

59. When using liquidtight flexible metal conduit (LFMC) to enclose service-entrance conductors and the equipment bonding jumper is routed with the LFMC, the conduit may be a MAXIMUM length of _____.

    a. 24 inches

    b. 48 inches

    c. 6 feet

    d. 10 feet

Notes

60. If two driven ground rods form the entire grounding electrode system of a temporary service, what is the MAXIMUM size copper conductor required to bond the ground rods together regardless of the size of the service?

    a. 8 AWG

    b. 6 AWG

    c. 4 AWG

    d. 2 AWG

61. Under which, if any, of the following conditions is the grounded (neutral) conductor to be considered as a current-carrying conductor?

    I. When it is only carrying the unbalanced current of a single-phase, 120/240 volt, three-wire electrical system.

    II. When it is the neutral of a three-phase, wye-connected electrical system that consists of nonlinear loads.

    a. I only

    b. II only

    c. both I and II

    d. neither I nor II

62. Luminaires (lighting fixtures) using mercury vapor or metal halide lamps installed in an indoor sports arena shall be _____.

    a. cord-and-plug-connected

    b. stem-mounted with rigid metal conduit

    c. prohibited

    d. protected with a glass or plastic lens

63. In general, which of the following panelboards, if any, is/are required to have the grounded conductor and the grounding conductor bonded together at the panelboard?

    I. main panelboard (used as service equipment)

    II. sub-panelboard

        a. I only

        b. II only

        c. both I and II

        d. neither I nor II

64. Border lights and footlights in a theater shall be arranged so that no branch circuit supplying such equipment carries a load exceeding _____.

        a. 15 amperes

        b. 16 amperes

        c. 20 amperes

        d. 30 amperes

65. Where rigid nonmetallic conduit (RNC) is buried underground about a bulk storage facility, the conduit is required to have NOT less than _____ of cover.

        a. 1½ feet

        b. 2 feet

        c. 3 feet

        d. 4 feet

66. The largest standard rating for an inverse time circuit breaker is _____.

        a. 600 amperes

        b. 1,200 amperes

        c. 6,000 amperes

        d. 8,000 amperes

Notes

67. What is the MINIMUM height allowed for a fence enclosing an outdoor installation of 2,400 volt electrical equipment?

    a. 7 feet

    b. 8 feet

    c. 9 feet

    d. 10 feet

68. What is the MAXIMUM number of current-carrying conductors permitted in a metal wireway without derating the ampacity of the conductors?

    a. 3

    b. 20

    c. 30

    d. 42

69. Determine the MAXIMUM number of size 4/0 AWG copper conductors with THWN insulation permitted to be installed in a 4 inch $\times$ 4 inch metal wireway.

    a. 20

    b. 12

    c. 10

    d. 9

70. Which of the following conductors, if any, are permitted by the NEC to be installed in the same raceway with the service-entrance conductors?

    a. grounding conductors

    b. sub-panel feeders

    c. branch-circuit conductors

    d. none of these

71. An aluminum grounding electrode shall be _____.

    a. installed at least 18 inches above the earth

    b. installed at least 24 inches above the earth

    c. in direct earth burial

    d. prohibited

72. Service conductors that supply a building are permitted to pass through the interior of another building _____.

    a. when the conductors are installed in rigid metal conduit (RMC)

    b. when an equipment grounding conductor is routed with the conductors

    c. if there are not more than three (3) current-carrying conductors in the raceway

    d. never

73. A switched, 15 ampere, 120 volt receptacle outlet located in the bedroom of a dwelling unit is _____.

    a. required to have GFCI protection

    b. required to have AFCI protection

    c. prohibited

    d. not permitted to have AFCI protection because the arcing characteristics of the switch would trip the AFCI device

74. For service installations consisting of not more than two (2), two-wire branch circuits, the service disconnecting means shall have a rating of NOT less than _____.

    a. 15 amperes

    b. 30 amperes

    c. 60 amperes

    d. 75 amperes

Notes

Notes

75. Lighting track shall NOT be installed _____.

    I.   within 3 feet horizontally of a shower or bathtub enclosure

    II. under an outdoor canopy

        a.   I only

        b. II only

        c. both I and II

        d. neither I nor II

Notes

ANSWER KEY & NEC® REFERENCES

Maintenance Electrician
Practice Exam #1 Answer Key

ANSWER	NEC PAGE #	REFERENCE
1. a	23	90.1(B)&(C)
2. d	100	Article 100
3. a	30	Article 100
4. c	47	210.8(B)(4)
5. a		General Knowledge
6. c		Trade Knowledge
7. b	198	358.26
8. d	310	Table 430.248
9. a	250	408.36(C)
10. c		Trade Knowledge
11. c	37	110.26(E)
12. d	133	310.3
13. c	250	408.35
14. c	83	240.6(A)
15. b	163	314.16(A)
16. d	201	362.30(A)
17. d	50	210.21(B)(1)
18. c	285	430.6(A)(1)
	310	Table 430.248
	290	430.22(A)
		Full-Load Current of 7½ HP motor
		FLC = 40 amperes $\times$ 125% = 50 amperes
19. a	142	Table 310.16
20. c	88	240.51(B)
	89	240.60(D)

<u>ANSWER</u>	<u>NEC PAGE #</u>	<u>REFERENCE</u>

21. c

Current formula $I = P/E$

$$I = \frac{10,000 \text{ watts}}{240 \text{ volts}}$$

$I = 41.6$ amperes

22. a   26   Article 100

23. a   26   Article 100

24. b

Current Formula $I = P/E$

150 watts $\times$ 15 luminaires = 2250 watts

$$I = \frac{2250 \text{ watts}}{120 \text{ volts}}$$

$I = 18.75$ amperes

25. d   243   404.8(A)

Maintenance Electrician
Practice Exam #2 Answer Key

ANSWER	**NEC PAGE #**	**REFERENCE**
1. d	59	210.18(B)
2. d	301	430.83(A)(2)
3. a	23	90.1(B)
4. b	27	Trade Knowledge
5. b	269	422.51
6. c		Trade Knowledge
7. d	35	110.15
8. b	659	Annex C, Table C.1
9. a	140	Table 310.15(B)(2)(a)
10. b	130	300.20(A)
11. a	198	358.30(A)
12. a		General Knowledge
13. c	286	430.6(A)(2)
14. a	290	430.22(A)
15. b	35	110.14(C)(1)(b)(1)
16. b	89	240.83(D)
17. d	136	310.12(C)
18. a	44	200.7(A)(2)
19. c		Three-Phase Current Formula

$$I = \frac{P}{E \times 1.732}$$

$$I = \frac{9,000 \text{ watts}}{208 \text{ volts} \times 1.732} = \frac{9,000}{360.25}$$

$$I = 24.9 \text{ amperes}$$

<u>ANSWER</u>	<u>NEC PAGE #</u>	<u>REFERENCE</u>

20. b   635   Chapter 9, Table 8
Single-Phase Voltage Drop Distance Formula

1. Find Allowable Voltage Drop
   VD = 240 volts $\times$ 3%
   VD = 7.2 volts

2. $D = \dfrac{CM \times VD}{2 \times K \times I}$

   $D = \dfrac{16,510 \times 7.2}{2 \times 12.9 \times 42}$

   $D = \dfrac{118,872}{1,083.6}$

   $D = 109.7$ ft.

21. b   187   348.20(A)(2)c.

22. d   37   110.26(A)(2)

23. a   49   210.19(A)(1)

24. c   120   285.3(1)

25. b   82   240.4(D)

Residential Electrician
Practice Exam #3 Answer Key

ANSWER	NEC PAGE #	REFERENCE
1. a	104	250.64(E)
2. b	245	404.14(E)
3. a	53	210.52(E)
4. d	242	Table 402.5
5. a	52	210.50(C)
6. d	79	230.90

7. c   265   422.11(E)(3)
           83   240.6(A)

Current Formula $I = P/E$

$$I = \frac{4500 \text{ volt-amperes}}{240 \text{ volts}}$$

$I = 18.75$ amperes

18.75 amperes $\times$ 150% = 28.1 amperes

*NOTE: The next size breaker is rated @ 30 amperes.

8. d	158	312.2(A)
9. a	447	550.10(A)

10. c   Trade Knowledge

Current Formula $I = P/E$

$$I = \frac{150{,}000 \text{ volt-amperes}}{240 \text{ volts}}$$

$I = 625$ amperes

11. c	54	210.63
12. c	58	Table 220.12

70 ft. $\times$ 30 ft = 2100 sq. ft.
2100 sq. ft. $\times$ 3 VA = 6300 VA (house)
120 volts $\times$ 15 amperes = 1800 VA (circuit)

$$\frac{6300 \text{ VA (house)}}{1800 \text{ VA (circuit)}} = 3.5 \text{ or } 4 \text{ branch circuits}$$

ANSWER	**NEC PAGE #**	**REFERENCE**
13. c	50	Table 210.21(B)(3)
14. d	73	230.9(A)
15. c		Trade Knowledge

Current Formula $I = P/E$

100 watts $\times$ 6 luminaires = 600 watts

$$I = \frac{600 \text{ watts}}{120 \text{ volts}}$$

$I = 5$ amperes

16. a	125	Table 300.5, Note 5
17. c	167	314.29
18. c	166	314.24
19. b	158	312.2(A)
20. a	53	210.52(H)
21. a	52	210.52(C)(1)
22. b	125	Table 300.5, Column 4
23. c	179	334.15(C)
24. d	136	310.12(C)
	43	200.6(A)-(E)
	111	250.119
25. b	242	404.2(A)

Residential Electrician
Practice Exam #4 Answer Key

ANSWER	NEC PAGE #	REFERENCE
1. c	53	210.52(C)(5)
2. d	91	240.24(E)
3. a	35	110.14(A)
4. c	83	240.5 (B)(2)
5. b	47	210.8(A)(6)
6. a	257	410.18(B), Ex.2
7. a	532	680.22(A)(1)
8. d	246	406.3(C)
9. b	50	210.23(A)(1)
		20 amperes $\times$ 80% = 16 amperes
10. a	164	314.20
11. c	78	230.79(C)
12. b	62	Table 220.55, Column B
		8 kW (connected load) $\times$ 80% = 6.4 kW demand load
13. c	454	550.32(C)
14. a	533	680.23(A)(3)
15. b	104	Table 250.66
16. c	481	590.3(B)
17. b	162	314.16(B)(3)
18. c	270	424.3(A)
19. b	198	358.30(A)
20. d	61	Table 220.54
21. a	183	340.80

ANSWER	NEC PAGE #	REFERENCE
22. a	164	314.20
23. b	47	210.8(A)(7)
24. a	52	210.52(B)(2)
	48	210.11(C)(2)&(3)
25. c	101	250.52(A)(1)

Residential Electrician
Practice Exam #5 Answer Key

ANSWER	NEC PAGE #	REFERENCE
1. d	267	422.16(B)(2)(1)
2. b	141	Table 310.15(B)(6)
3. c	166	314.27(D)
4. b	102	250.53(G)
5. b	180 142	334.80 Table 310.16
6. c	163	Table 314.16(A)
7. c	57 & 58 58	220.12 Table 220.12 2600 sq. ft. $\times$ 3 VA = 7800 VA (total lighting) 120 volts $\times$ 15 amperes = 1800 VA (one circuit) $\dfrac{7800\ VA}{1800\ VA} = 4.3$ or 5 circuits
8. a	59	220.14(J)(1)
9. d	60	220.52(A) 1,500 VA $\times$ 2 circuits = 3,000 VA
10. b	60	220.53
11. d	53	210.52(D)
12. c	124	300.4(D)
13. b		Power Formula $\quad P = I \times E$ $P = 30$ amperes $\times$ 240 volts = 7,200 VA
14. d	139 134	Table 310.13 310.8(B)&(C)(2)
15. b	162	314.16(B)(4)
16. c	60	220.54

ANSWER	**NEC PAGE #**	**REFERENCE**
17. c	51	210.23(B)
		30 amperes × 80% = 24 amperes
18. d	246	406.3(A)
19. d	62	Table 220.55 & Note 3
		Use column B
		Two Appliances = 65% demand
		5 kW + 7 kW = 12 kW (total connected load)
		12 kW × 65% = 7.8 kW demand load
20. a	112	Table 250.122
21. a	46	210.6(A)(1)
22. d	270	424.3(B)
23. c	268	422.35
24. b	261	410.66(A)(1)
25. d	539	680.43(B)(1)(a)

Residential Electrician
Practice Exam #6 Answer Key

ANSWER	NEC PAGE #	REFERENCE
1. d	99	250.32(A), Ex.
2. b	74	230.26
3. d	305	430.109(B)
4. c	74	230.28
5. a	539	680.43(C)
6. c	532	680.22(A)(5)
7. d	166	314.27(B)
8. a	95	250.8
9. c	102	250.53(G)
10. d	247	406.8(C)
11. c	247	406.4(E)
12. a	52	210.52(C)(1)
13. b	278	426.4
14. b	125	Table 300.5, Column C
15. c	54	210.70(A)(2)(c)
16. d	76	230.54(C), Ex.
17. a	129	300.16(A)
18. c	530	680.9
19. b	224	394.10(1)
20. d	47	210.8(A)(7)
21. b	180	334.104
22. c	163	Table 314.16(B)
23. b	180	334.30

ANSWER	NEC PAGE #	REFERENCE
24. a	74	230.31(B)
25. b	163	Table 314.16(B)
	162	314.16(B)(4)

$$\frac{18 \text{ cu. in.(box)}}{2.25 \text{ cu. in.(\#12 wire)}} = 8 \text{ wires}$$

8 wires (allowable fill)
-2 wires (device)
———————————————
6 wires may be added

Size 12/2 with ground type NM cable = 3 wires in cable

$$\frac{6 \text{ wires (allowed)}}{3 \text{ wires (per cable)}} = 2 \text{ cables are permitted}$$

Journeyman Electrician
Practice Exam #7 Answer Key

ANSWER	NEC PAGE #	REFERENCE
1. a	265	422.11(F)(1)
2. d	46	210.5(C)
3. d	437	525.32(B)(4)
4. b	103	250.53(H)
5. d	142 140	Table 310.16 Table 310.15(B)(2)(a) Ampacity of 1/0 THW Copper = 150 amperes 150 amperes × 80% (adjustment factor) = 120 amperes
6. a	27	Article 100
7. c	290 142	430.22(A) Table 310.16 FLC of Motor = 70 amperes × 125% = 87.5 amperes Size 3 AWG THW ampacity = 100 amperes
8. c	82 83	240.4(B)(2)&(3) 240.6(A)
9. b	311	Table 430.250
10. d	108	250.104(B)
11. b	659	Annex C, Table C.1
12. b	55	215.3 240 amperes × 125% = 300 amperes
13. c	189	350.30(A)
14. c	166	314.28(A)(1)
15. c	481 230	590.4(B) Table 400.4
16. c	188	Table 348.22

ANSWER	NEC PAGE #	REFERENCE
17. d	533	680.23(B)(2)
18. c	531	680.10
19. a	565	700.12(A)
20. c	78	230.79(D)
21. b	54	210.60(B)
22. b	259	410.33
23. c	124	300.4(F)
24. c	343	500.5(D)
25. a		Current Formula $I = P/E$

$$I = \frac{15{,}000 \text{ VA}}{240 \text{ volts}} = 62.5 \text{ amperes}$$

Journeyman Electrician
Practice Exam #8 Answer Key

ANSWER	NEC PAGE #	REFERENCE
1. a	300	430.74(A)
2. d	249	408.7
3. d	58	220.14(I)
4. d	138 142	Table 310.13 Table 310.16
5. a	101	250.50, Ex.
6. a	422	517.64(A)(1),(2)&(3)
7. a	143	Table 310.17
8. c	162	314.16(4)
9. d	285	430.6(A)(1)
10. a	58	220.14(F)
11. a	707	Annex C, Table C.10(A)
12. b	37	110.26(E)
13. a	342	500.5(B)
14. d	127	300.6(D)
15. c	45	210.3
16. d	538	680.41
17. b	516	645.10
18. c	178	332.30
19. a	166	314.24
20. a	124	300.5(D)(1)
21. b	112	Table 250.122
22. c	79	230.95

ANSWER	NEC PAGE #	REFERENCE
23. a	142	Table 310.16
	140	Table 310.15(B)(2)(a)

Ampacity of size 500 kcmil THWN Copper Conductors before adjustment and correction factors = 380 amperes

380 amperes $\times$ .67 (temperature correction) $\times$ .7 (adjustment factor) = 178.22 amperes

24. c	466	551.77(A)
25. c	351	501.15(C)(3)

Journeyman Electrician
Practice Exam #9 Answer Key

ANSWER	NEC PAGE #	REFERENCE
1. a	37	110.26(F)(1)(a)
2. b	175	328.10(3)
	218	392.3(B)(1)(a)
3. c	311	Table 430.250
	296	Table 430.52
	295	430.52(C)(1), Ex.1

FLC of Motor = 4.6 amperes $\times$ 300% = 13.8 amperes

*NOTE: The next standard size fuse is rated at 15 amperes.

4. b	364	503.5
5. d	210	376.56
6. a	191	352.26
7. b	102	250.52(A)(5)(a)
8. d	189	350.12(1)
9. b	35	110.14(C)
	136	310.15(A)(2)
	142	Table 310.16

Size 3 AWG THHN ampacity before derating
= 110 amperes (from 90° C column)
110 amperes $\times$ .96 (temperature correction)
= 105.6 amperes

Size 3 AWG THHN (from 60° C column)
= 85 amperes is to be used

10. a	77	230.71(A)
11. b	191	Table 352.30(B)
12. b	296	Table 430.52, Note 1
13. b	218	392.3(B)(1)(a)

ANSWER	NEC PAGE #	REFERENCE
14. b	319	440.64
15. a	47	210.8(B)(2)
16. a	392	513.8(A)
17. d	291	430.24
18. a	112	Table 250.122
19. b	304	430.102(B), Ex.
20. b	359	502.10(A)(4)
21. a	104	Table 250.66

Size 500 kcmil $\times$ 4 conductors
= 2000 kcmil total

Aluminum service-entrance conductors over 1750 kcmil require a size 3/0 AWG copper grounding electrode conductor.

22. b	508	Table 630.11(A)
	142	Table 310.16

.78 (duty cycle) x 50 amperes (primary current)
= 39 amperes
39 amperes requires size 8 AWG copper conductors.

23. b	516	647.1
24. b	412	517.19(B)(1)&(2)
25. d	128	300.13(B)

Journeyman Electrician
Practice Exam #10 Answer Key

ANSWER	NEC PAGE #	REFERENCE
1. d	249	408.5
2. b	121	285.21(A)(1)
3. c	102	250.52(A)(3)
4. c	127	300.7(A)
5. c	88	240.40
6. d	635	Chapter 9, Table 8

Single-Phase Voltage Drop Formula

$$\text{Voltage Drop} = \frac{2 \times K \times I \times D}{CM}$$

$$VD = \frac{2 \times 12.9 \times 80 \text{ amperes} \times 200 \text{ ft.}}{52,620 \text{ CM}}$$

$$VD = \frac{412,800}{52,620}$$

$$VD = 7.84 \text{ volts}$$

7. b	350	501.15(A)(1)
8. d	255	410.8(D)(1)
9. d	395	Table 514.3(B)(1)
10. b	415	517.31
11. a	582	725.82(H)
12. b		Trade & General Knowledge

Three-Phase Current Formula
$$I = P/(E \times 1.732)$$

$$I = \frac{90,000 \text{ VA}}{208 \text{ volts} \times 1.732}$$

$$I = \frac{90,000}{360.25}$$

$$I = 250 \text{ amperes}$$

ANSWER	NEC PAGE #	REFERENCE
13. c	537	680.26(C)
14. d	182	338.10(A)
15. c	565	700.12(A)
16. c	51	Table 210.24
17. d	294	430.36
18. b	163	Table 314.16(B)

Size 10 AWG wire = 2.5 cubic inches
Size 12 AWG wire = 2.25 cubic inches
2.25 cu. in. $\times$ 6 (wires in box) = 13.5 cu. in.

$$\begin{array}{r} 27.0 \text{ cubic inches (box)} \\ \underline{-13.5 \text{ cubic inches (existing wire in box)}} \\ 13.5 \text{ cubic inches (remaining space)} \end{array}$$

$$\frac{13.5 \text{ cu. in. (remaining)}}{2.5 \text{ cu. in. (size 10 AWG)}} = 5.4 \text{ or 5 wires}$$

ANSWER	NEC PAGE #	REFERENCE
19. b	448	550.10(G)
20. d	390	511.7(A)(1)
21. d	343	500.5(C)
22. c	265	422.10(A)
	266	422.13(C)
23. a	265	422.10(B)
24. d	102 & 103	250.53(G)
25. c	627	Chapter 9, Table 4
	631	Chapter 9, Table 5

Size 1 AWG THW = .1901 sq. in. $\times$ 5 wires = .9505 sq. in.
Size 3 AWG THW = .1134 sq. in. $\times$ 5 wires = .567 sq. in.
.9505 sq. in. + .567 sq. in. = 1.5175 sq. in.

*NOTE: 2½" IMC at 40% Fill = 2.054 sq. in.

Journeyman Electrician
Practice Exam #11 Answer Key

<u>ANSWER</u>	<u>NEC PAGE #</u>	<u>REFERENCE</u>
1. d	35	110.14(C)(1)(a)(1)
2. d	107 104	250.102(C) 250.66 2,000 kcmil $\times$ 12.5% = 250 kcmil
3. d	92 105	250.4(A)(5) 250.54
4. a	300	430.72(C)(3)
5. c	272 272 273	424.38(B)(1) 424.34 424.41(H)
6. c	162 163	314.16(B)(1) Table 314.16(B) 2.25 cu. in. $\times$ 2 wires = 4.50 cu. in.
7. a	179	334.10(2)
8. b	480 389	555.22 511.3(B)(2)
9. c	242	Table 402.5
10. c	412	517.18(C)
11. b	632 626	Chapter 9, Table 5 Chapter 9, Table 4 Size 2 AWG THWN = .1158 sq. in. $\times$ 3 wires = .3474 sq. in. Size 6 AWG THWN = .0507 sq. in. $\times$ 4 wires = .2028 sq. in. Size 10 AWG THWN = .0211 sq. in. $\times$ 3 wires = .0633 sq. in. Total = .6135 sq. in. *NOTE: 1½" EMT at 40% Fill = .814 sq. in.
12. d	184	340.112

ANSWER	**NEC PAGE #**	**REFERENCE**
13. a	170	320.80(A)
14. d	51	Table 210.24
15. b	169	320.17
	176	330.17
16. c	55	215.3
17. c	120	285.3(1)
18. b	125	Table 300.5
19. d	212	382.15
20. c	255	410.8(B)&(C)
21. a	58	220.14(I)

200 receptacles $\times$ 180 VA = 36,000 VA

22. d	250	408.34(E)
23. a	531	Table 680.8
24. d	58	Table 220.12
	27	Article 100
	49	210.19(A)(1)

1,600 sq. ft. $\times$ 3 VA = 4,800 VA
4800 VA $\times$ 125% = 6,000 VA (demand load)

120 volts $\times$ 15 amperes = 1,800 VA (one circuit)

$$\frac{6,000 \text{ VA (demand load)}}{1,800 \text{ VA (one circuit)}} = 3.3 \text{ or 4 circuits}$$

| 25. d | 102 | 250.52(B)(2) |

Master Electrician
Practice Exam #12 Answer Key

ANSWER	NEC PAGE #	REFERENCE
1. c	98 104	250.30(A)(3) Table 250.66 300 kcmil × 3 conductors = 900 kcmil *NOTE: Table 250.66 requires a size 2/0 AWG grdg. conductor.
2. b	126	300.5(K)
3. b	610	820.15
4. c	305	430.109(C)(2) 15 amperes × 80% = 12 amperes
5. d	562	695.7 480 volts × 85% = 408 volts
6. b	37	110.26(D)
7. a	87	240.24(B), Ex.1(1)
8. b	142 140	Table 310.16 Table 310.15(B)(2)(a) Size 250 kcmil THWN copper conductor ampacity before corrections and adjustments = 255 amperes 255 amperes × .82 (temperature correction) = 209.1 amperes 209.1 amperes × .8 (adjustment factor) = 167.28 amperes
9. d	299	Table 430.72(B)
10. c	102 108	250.52(B)(1) 250.104(B)
11. b	102	250.52(A)(4)
12. d	551	690.31(B)
13. d	137	Table 310.13
14. b	250	408.35 *NOTE: The physical arrangement of the busbars does not permit more than 10 double-pole breakers to be installed on each side of the panel.

ANSWER	**NEC PAGE #**	**REFERENCE**

15. b   58   Table 220.12

12,000 sq. ft. $\times$ 3 VA = 36,000 VA (building)
120 volts $\times$ 15 amperes = 1800 VA (one circuit)

$$\frac{36,000 \text{ VA (building)}}{1800 \text{ VA (one circuit)}} = 20 \text{ circuits}$$

16. d   54   210.62

17. a   72   230.6(1)

18. b   292   430.32(A)(1)

19. d   635   Chapter 9, Table 8
Single-Phase Voltage Drop Formula

$$VD = \frac{2 \times K \times I \times D}{CM}$$

$$VD = \frac{2 \times 12.9 \times 90 \text{ amperes} \times 225 \text{ ft.}}{52,620 \text{ CM}}$$

$$VD = \frac{522,450}{52,600} \qquad VD = 9.92 \text{ volts}$$

20. c   530   Figure 680.8
                    531   Table 680.8

21. b   397   Table 514.3(B)(2)

22. c   390   511.12

23. c   311   Table 430.250
                    291   430.24

FLC of 10 HP motor = 30.8 amperes
30.8 amperes $\times$ 125% = 38.5 amperes
FLC of 7½ HP motor = 24.2 amperes
24.2 amperes $\times$ 100% = 24.2 amperes
38.5 amperes + 24.2 amperes = 62.7 amperes

24. c   249   Table 408.5

25. c   107   250.102(D)
                    111   250.122(C)
                    112   Table 250.122

Master Electrician
Practice Exam #13 Answer Key

<u>ANSWER</u>	<u>NEC PAGE #</u>	<u>REFERENCE</u>
1. d	541	680.57(C)(1)
2. d	29	Article 100
3. c	441	540.13
4. a	394	514.8, Ex.2
5. b	85	240.21(B) *NOTE: The next larger standard size overcurrent
	82	240.4(B) device is not permitted for feeder taps.

6. b   162   314.16(B)(2),(4)&(5)

2 clamps	= 1 conductor
1 device	= 2 conductors
3 grdg. cond.	= 1 conductor
Total	= 4 conductors

7. a	27	Article 100
8. c	541	680.58
9. d	397	514.11(B)
10. c	423	517.71(A), Ex.
11. d	311	Table 430.250
	290	430.22(A)
	142	Table 310.16

FLC of 25 HP motor = 74.8 amperes $\times$ 125% = 93.5 amperes
*NOTE: The wire size needs to be increased because of the
 temperature correction factor.
Size 1 THWN ampacity = 130 amperes
130 amperes $\times$ .75 (temperature correction) = 97.5 amperes
<u>OR</u> 93.5 amperes $\div$ .75 (temp. correction) = 124.6 amperes

12. d	100	250.34(A)(2)
13. b	416	517.33(A)(5)
14. b	444	547.9(A)(4),(5),(6)&(7)

ANSWER	**NEC PAGE #**	**REFERENCE**
15. c	66	Table 220.103
16. a	409	620.22(A)
17. b	465	551.71
18. d	527 & 528	675.11(A)&(C)
19. b	323	Table 450.3(B)

Three-Phase Current Formula

$$I = \frac{VA}{E \times 1.732} \qquad I = \frac{25,000\ VA}{208 \times 1.732}$$

$$I = \frac{25,000}{360.25} \qquad I = 69.3\ \text{amperes} \times 125\% = 86.62\ \text{amperes}$$

20. b	112	250.122(F)
21. a	445	547.9(D)
22. c	311	Table 430.250
	291	430.24
	142	Table 310.16

40 HP FLC = 52 amperes × 100% = 52 amperes
50 HP FLC = 65 amperes × 100% = 65 amperes
60 HP FLC = 77 amperes × 125% = 96 amperes
Total = 213 amperes

*NOTE: Size 4/0 THWN copper ampacity = 230 amperes

23. d	297	430.53(A)(1)
24. a	389	511.3(A)(7)
25. c	635	Chapter 9, Table 8

Three-Phase Voltage Drop Conductor Size Formula
*NOTE: Voltage drop permitted is 3%
480 volts × 3% = 14.4 volts (permitted drop)

$$CM = \frac{1.732 \times K \times I \times D}{VD}$$

$$CM = \frac{1.732 \times 21.2 \times 100\ \text{amperes} \times 390\ \text{ft.}}{14.4} = 99,446\ CM$$

Size 1/0 AWG Aluminum = 105,600 CM

Master Electrician
Practice Exam #14 Answer Key

ANSWER	NEC PAGE #	REFERENCE
1. b	571	702.10(B)
2. a	75	230.44, EX.
3. d	560	695.5(B)
4. b	54	210.60(B)
5. d	411 & 412	517.18(A)&(B)
6. b	96	250.24(C)(1)
	104	Table 250.66
7. c	49	210.19(A)(1)
	48	210.11(A)

400 kVA = 400,000 VA

400,000 VA $\times$ 125% (continuous load)

= 500,000 VA (building lighting)

277 volts $\times$ 20 amperes = 5540 VA (one circuit)

$$\frac{500,000 \text{ VA (bldg. lighting)}}{5540 \text{ VA (one circuit)}} = 90.2 \text{ or } 91 \text{ circuits}$$

*NOTE: Circuits need only to be installed to serve the connected load.

8. b	243	404.8(B)
9. a	127	300.7(A)
10. d	411	517.13(B), Ex.2
11. c	533	680.23(B)(2)(b)
12. b	320	445.13
13. a	59	220.14(K)(2)
14. d	303	Table 430.91
15. c	562	695.14(D)

<u>ANSWER</u>	<u>NEC PAGE #</u>	<u>REFERENCE</u>

16. c   142   Table 310.16
             140   Table 310.15(B)(2)(a)

Size 750 kcmil XHHW aluminum ampacity
before derating = 435 amperes
435 amperes $\times$ 1.04 (temperature correction)
$\times$ .8 (adjustment factor) = 361.92 amperes

17. c   219   392.6(G)&(H)

18. c   172   324.10(A)

19. c   142   Table 310.16
                       Current Formula   I = P/E

$$I = \frac{23,600 \text{ VA}}{240 \text{ volts}}$$

I = 98.33 amperes

* NOTE: Size 1 AWG USE aluminum cable has
  an ampacity of 100 amperes.

20. b   530   680.8(C)
             531   Table 680.8

21. d   60   Table 220.42

205.4 kVA = 205,400 VA total lighting
The first 3,000 VA at 100%   =   3,000 VA
The next 117,000 VA at 35%   = 40,950 VA
The remaining 85,400 VA at 25% = 21,350 VA
                                      Total = 65,300 VA or 65.3 kVA

22. a   631 & 632   Chapter 9, Table 5
             626   Chapter 9, Table 4

Size 10 AWG TW   = .0243 sq. in. $\times$ 24 = .5832 sq. in.
Size 10 AWG THW   = .0333 sq. in. $\times$ 10 = .3330 sq. in.
Size 12 AWG THHN = .0133 sq. in. $\times$ 14 = .1862 sq. in.
                                      Total = 1.1024 sq. in.

Trade size 2" EMT at 40% Fill = 1.3242 sq. in.

ANSWER	NEC PAGE #	REFERENCE
23. a	483	600.2
	483	600.3(A)
24. d	316	440.22(A)
25. d	58	220.14(H)(2)

90 ft. $\times$ 180 VA per ft. = 16,200 VA (multioutlet assembly)

20 amperes $\times$ 120 volts = 2400 VA (one circuit)

$$\frac{16,200 \text{ VA (total load)}}{2400 \text{ VA (one circuit)}} = 6.75 \text{ or } 7 \text{ circuits}$$

Notes

Master Electrician
Practice Exam #15 Answer Key

<u>ANSWER</u>	<u>NEC PAGE #</u>	<u>REFERENCE</u>
1. a	569	701.11
2. a	123	300.3(C)(1)
3. a	565	700.12
4. c	132	Table 300.50
5. a	502	620.61(B)(1)
6. d	121	285.25
7. c	465	551.71
8. a	285	430.6(A)(1)
	311	Table 430.250
	290	430.22(A)
	236	Table 400.5(A), Column A
		FLC of 40 HP motor = 52 amperes $\times$ 125% = 65 amperes
9. b	311	Table 430.250
		63 amperes $\times$ 1.1 (power factor) = 69.3 amperes
10. a	62	Table 220.55 & Note 3
		Use column B 5 appliances at 45% demand 6 kW + 8 kW + 3.5 kW + 6 kW + 3.5 kW = 27 kW (total connected load) 27 kW $\times$ 45% = 12.15 kW (demand load)
11. d	566	700.12(E)
12. c	204	368.17(B), Ex.
13. c	89	240.60(C)(3)
14. c	465	551.71 40 sites $\times$ 20% = 8 sites
15. d	210	376.30(A)
16. a	125	Table 300.5

ANSWER	**NEC PAGE #**	**REFERENCE**

17. a   286   430.6(A)(2)
                292   430.32(A)(1)

FLA of motor = 18 amperes $\times$ 115% = 20.7 amperes

18. d   293   430.32(C)

FLA of motor = 18 amperes $\times$ 130% = 23.4 amperes

19. b   321   450.3(B)
                323   Table 450.3(B)
                83   240.6(A)

Three-Phase Current Formula $I = \dfrac{P}{E \times 1.732}$

$I = \dfrac{50,000 \text{ VA}}{480 \times 1.732} = \dfrac{50,000}{831.36}$

I = 60.2 amperes $\times$ 250% = 150.5 amperes

20. a   59   220.44
                58   220.14(I)
                60   Table 220.44   150 receptacles $\times$ 180 VA = 27,000 VA

(First) 10,000 VA at 100% = 10,000 VA
(remainder) 17,000 VA at 50% = 8,500 VA
10,000 VA + 8,500 VA = 18,500 VA total demand load

21. a   162   314.16(B)(4)
                163   Table 314.16(A)

3½" masonry box = 9 conductors (#12 AWG)
are permitted per gang
1 switch = 2 conductors (deduct per box)
9 − 2 = 7 conductors per box $\times$ 3 gang = 21 conductors

22. c   359   502.15(2)

23. b   99   250.30(A)(7)(2)

24. a   444   547.5(F)

25. c   325   450.13(B)

Sign Electrician
Practice Exam #16 Answer Key

ANSWER	**NEC PAGE #**	**REFERENCE**
1. b	134 30	310.8(B)&(C)(2) Article 100
2. a	110	250.118(2),(3)&(4)
3. b	128	300.14
4. b	49	210.19(A)(1), FPN# 4 120 volts $\times$ 3% = 3.6 volts permitted drop
5. b		Trade Knowledge
6. d	486	600.32(A)(5)
7. d	125	Table 300.5
8. a	485	600.10(C)(2)
9. b		Trade Knowledge
10. d	486	600.31(B)
11. a	703	ANNEX C, Table C.10
12. a	541	680.57(C)(2)
13. d	541	680.57(C)(1)
14. c	484	600.9(A)
15. b	483	600.5(B)(1)
16. a	95	250.8
17. c	78	230.79(D)
18. b	259	410.33
19. a	133	310.3
20. b	484	600.6(B)
21. d	27	Article 100

ANSWER	NEC PAGE #	REFERENCE
22. a	67	225.7(C)
23. b	142	Table 310.16
24. b	483	600.5(A)
25. d	484	600.6(A)(1)
	29	Article 100

Sign Electrician
Practice Exam #17 Answer Key

<u>ANSWER</u>	<u>NEC PG.#</u>	<u>REFERENCE</u>
1. b	485	600.21(E)
2. c	59	220.18(B)
3. a		Trade Knowledge
4. a	483	600.2
	483	600.3(A)
5. b	541	680.57(A)&(B)
6. c	34	110.9
7. c	163	Table 314.16(B)
	163	Table 314.16(A)
		Size 12 AWG = 2.25 cu. in. $\times$ 2 = 4.5 cu. in.
		Size 8 AWG = 3.0 cu. in. $\times$ 3 = 9.0 cu. in.
		4.5 cu. in. + 9.0 cu. in. = 13.5 cu. in.
8. c	485	600.10(D)(2)
9. c	82	240.4(D)
10. a	484	600.7(D)
11. d	484	600.8(C)
12. a	188	348.30(A)
13. d	191	352.30(A)
14. c	191	352.26
15. c	483	600.5(B)(2)
16. b	69	225.25(2)
17. a	483	600.5(B)(3)
	256	410.15(B)(1)
18. b	67	225.11
	76	230.52

ANSWER	**NEC PAGE #**	**REFERENCE**
19. b	192	Table 352.44(A)
		3.65 inches $\times$ 2 = 7.3 inches
20. a	35	110.14(A)
21. a	673	Annex C, Table C.4
22. a		Trade Knowledge
23. c		Trade Knowledge
24. b	67	225.6(A)(1)
25. a	394	Figure 514.3

Residential Electrician
Final Exam Answer Key

ANSWER	NEC PAGE #	REFERENCE
1. c	29	Article 100
2. d	164	314.17(C)
3. a	257	410.18(B), Ex.2
4. b	62	Table 220.55, Note 1 & 4
	180	334.80
	142	Table 310.16

Current Formula $I = P/E$

17 kW $-$ 12 kW = 5 kW

5 kW $\times$ 5% = 25% (increase in column C)

8 kW (one appliance) $\times$ 125% = 10 kW (demand)

$$I = \frac{10{,}000 \text{ watts}}{240 \text{ volts}}$$

I = 41.66 amperes

Size 6 AWG NM cable ampacity = 55 amperes

ANSWER	NEC PAGE #	REFERENCE
5. c	265	422.11(E)(3)
6. a	164	314.23(B)(1)
7. d	180	334.104
8. c	162	314.16(B)(4)
9. b	78	230.79(C)
10. a	181	334.80
11. b	49	210.19(A)(3)
12. d	53	210.52(D)
13. a	533	680.22(C)
14. d	47	210.8(A)(7)
15. a	267	422.16(B)(4)
16. c	53	210.52(E)

ANSWER	**NEC PAGE #**	**REFERENCE**
17. b	124	300.5(D)(3)
18. a	44	200.7(C)(2)
19. b	47	210.8(A)(2), Ex.#2
20. d	99	250.32(A), Ex.
21. c	249	408.7
22. d	54	210.70(A)(1), Ex.#2
23. b	54 54	210.70(A)(1), Ex.#1 210.70(A)(2)(a)
24. b	164	314.17(C), Ex.
25. a	532	680.22(A)(1)
26. c	481 482	590.4(D) 590.6(A)
27. a	125	Table 300.5, Column 5
28. b	48	210.11(C)(1)
29. d	261 187	410.67(C) 348.20(A)(2)c.
30. a	319	440.62(B)
31. b	254	410.4(D)
32. b	123	300.4(B)(1)
33. a	532	680.21(A)(5)
34. c	481	590.3(B)
35. c	244	404.9(B), Ex.
36. b	102 & 103	250.53(G)
37. a	538 532	680.42 680.22(A)(3)
38. b	87	240.24(E)
39. a	48	210.11(C)(1),(2)&(3)

ANSWER	**NEC PAGE #**	**REFERENCE**

40. a   60   220.53

4800 VA + 1200 VA + 1150 VA
+ 800 VA + 1200 VA = 9150 VA total

9150 VA × 75% (demand) = 6862.5 VA

41. d   82   240.4(D)

42. c   141   Table 310.15(B)(6)

43. a   270   424.3(B)
            49   210.19(A)(1)

20 amperes ÷ 125% = 16 amperes   <u>OR</u>
20 amperes × 80%   = 16 amperes

44. c   54   210.63

45. b   531   680.21(A)(1)

46. d   53   210.52(G)

47. c   104   250.66
            104   Table 250.66

48. d   47   210.8(A)(2)

49. a   52   210.52

50. a   53   210.52(H)

51. b   162   314.16(B)(1),(2),(4)&(5)
            163   Table 314.16(B)

Size 14/2	= 2.00 cu. in. × 4 =	8.00 cu. in.
Size 12/2	= 2.25 cu. in. × 4 =	9.00 cu. in.
Equip. Grnd.	= 2.25 cu. in. × 1 =	2.25 cu. in.
Clamps	= 2.25 cu. in. × 1 =	2.25 cu. in.
Receptacle	= 2.25 cu. in. × 2 =	4.50 cu. in.
Switch	= 2.00 cu. in. × 2 =	4.00 cu. in.
	TOTAL =	30.00 cu. in.

*NOTE: One or more clamps are counted as equal to the largest wire in the box per 314.16(B)(2). One or more equipment grounding conductors are counted as equal to the largest equipment grounding conductor in the box per 314.16(B)(5). Devices are counted as equal to two conductors based on the largest conductor connected to the device per 314.16 (B)(4). All references per the NEC.

ANSWER	**NEC PAGE #**	**REFERENCE**
52. b	236	Table 400.5(A), Column B
53. c	179 & 180	334.15(C)
54. c	184	340.112
55. d	180	334.30(B)(1)&(2)
56. b	265	422.11(E)(3)
	83	240.6(A)

Current Formula $I = P/E$

$$I = \frac{5,000 \text{ watts}}{240 \text{ volts}}$$

I = 20.8 amperes $\times$ 150% = 31.2 amperes

*NOTE: The next standard size circuit breaker is rated at 35 amperes.

57. c	270	424.3(B)
	49	210.19(A)(1)

Power Formula $P = E \times I$

P = 240 volts $\times$ 20 amperes = 4,800 VA (one circuit)

1,000 watts $\times$ 125% = 1,250 VA (demand of one heater)

$$\frac{4,800 \text{ VA (one circuit)}}{1,250 \text{ VA (one heater)}} = 3.84 \text{ or 3 heaters}$$

58. d	534	680.23(B)(6)
59. a	531	680.12
60. a	164	314.21

Journeyman Electrician
Final Exam Answer Key

ANSWER	NEC PAGE #	REFERENCE
1. a	296	Table 430.52
	295	430.52(C)(1), Ex.1
2. c	44	210.7(A)(2)
3. c	140	310.15(B)(2), Ex.5
4. d	131	300.22(B)
5. c	397	514.9(A)
6. b		Trade Knowledge
7. d		Current Formula I = P/E

$$I = \frac{15,000 \text{ watts}}{208 \text{ volts}}$$

$$I = 72.1 \text{ amperes}$$

ANSWER	NEC PAGE #	REFERENCE
8. c	264	410.103
9. b	348	500.8(D)
10. b	320	445.14
11. b	35	110.14(C)(1)(b)(1)
12. a	250	408.34(A)
13. d	531	Table 680.8
14. c	74	230.26
15. d	126	300.6(A)
16. d	411	517.18(A)
17. b	489	605.8(C)
18. a	562	695.14(E)
19. c	543	680.74
20. d	237	400.14

ANSWER	**NEC PAGE #**	**REFERENCE**

21. c   305   430.110(A)

22. d   329   455.9

23. c   299   430.72(B)(2)
299   Table 430.72, Column B

24. c   291   430.24

25. a   222   392.10(A)(1)

26. a   58   220.14(I)
Current Formula   I = P/E

50 receptacles × 180 VA (each) = 9,000 VA (load)

$$I = \frac{9,000\ VA}{240\ volts}$$

I = 37.5 amperes

27. d   Power Formula   P = I × E

120 volts × 20 amperes = 2,400 VA (one circuit)

$$\frac{9,000\ VA\ (load)}{2,400\ VA\ (one\ circuit)} = 3.75\ or\ 4\ circuits$$

28. b   142   Table 310.16
140   Table 310.15(B)(2)(a)

Size 10 AWG THHN copper ampacity before derating = 40 amperes

40 amperes × 50% (adjustment factor for 20 wires) = 20 amperes

29. d   108   250.106, FPN# 2

30. d   117   250.178

31. c   215   386.56
216   388.56

32. d   576   725.26(B)(1)
123   300.3(C)(1)

33. d   29   Article 100

ANSWER	**NEC PAGE #**	**REFERENCE**
34. c	311	Table 430.250
	291	430.24

FLC of 10 HP motor = 30.8 amperes $\times$ 125% = 38.5 amperes
FLC of 7½ HP motor = 24.2 amperes $\times$ 100% = 24.2 amperes
38.5 amperes + 24.2 amperes = 62.7 amperes

35. b	130	300.20(A)
36. a	259	410.33
	138	Table 310.13
37. d	576	725.28(A)
38. d	247	406.8(C)
39. c	300	430.81(B)
40. c	104	250.66(A)
41. d	215	386.30
42. c	538	680.41
43. d	477	553.7(B)
44. c	465	551.71

50 sites $\times$ 70% = 35 sites

45. c	479	555.17(B)
46. d	186	344.30(B)(3)
47. a	206	370.3
48. b	166	314.24
49. a	96	250.24(A)(5)
50. b	103	250.56
51. a	561	695.6(A)
	72	230.6(2)
52. c	47	210.8(B)(4)
	247	406.8(B)(1)
53. b	593	770.133

ANSWER	NEC PAGE #	REFERENCE
54. b	54	210.62

$$\frac{36 \text{ ft. (length of show window)}}{12 \text{ ft. (length per receptacle)}} = 3 \text{ receptacles}$$

55. c	515	645.5(A)
56. c	101 104	250.52(A)(2)(1)&(3) 250.56
57. b	288	430.9(C)
58. c	186 187	344.30(B)(2) Table 344.30(B)(2)
59. a	142 140	Table 310.16 Table 310.15(B)(2)(a)

Size 4 AWG THW copper ampacity before derating = 85 amperes

85 amperes $\times$ .88 (temperature correction) $\times$ .8 (adjustment factor) = 59.84 amperes

60. a	201	362.30(A)
61. d	162 163	314.16(B)(1)&(5) Table 314.16(B)

Size 10 AWG = 2.50 cu. in. $\times$ 6 wires = 15.0 cu. in.
Size 12 AWG = 2.25 cu. in. $\times$ 10 wires = 22.5 cu. in.
Size 14 AWG = 2.00 cu. in. $\times$ 1 wire = 2.0 cu. in.
Total = 39.5 cu. in.

62. a	201	362.60
63. a	361	502.115(B)
64. b	270 49	424.3(B) 210.19(A)(1)

$$\frac{20 \text{ amperes}}{125\%} = 16 \text{ amperes}$$

65. c	305	430.110(A)

<u>ANSWER</u>	<u>NEC PAGE #</u>	<u>REFERENCE</u>
66. c	142	Table 310.16
	140	Table 310.15(B)(2)(a)
		Size 750 kcmil XHHW Aluminum ampacity (before derating) = 435 amperes 435 amperes $\times$ 1.04 (temperature correction) $\times$ .8 (adjustment factor) = 361.92 amperes
67. d	65	220.103
	66	Table 220.103
68. b	124	300.5(D)(1)
69. c	166	314.28(A)(1)
70. b	54	210.60(B)
71. b	264	410.104
72. a	111	250.122(B)
73. d	397	514.11(A)
74. c	443	547.5(A)
75. b	102	250.52(A)(5)

Notes

Master Electrician
Final Exam Answer Key

ANSWER	**NEC PAGE #**	**REFERENCE**
1. d	30 & 31	Article 100 Definition
	142	Table 310.16
		Current Formula I = P/E

$$I = \frac{90{,}000 \text{ VA}}{240 \text{ volts}} = 375 \text{ amperes}$$

*NOTE: Size 500 kcmil THHW copper with an ampacity of 380 amperes is required.

| 2. b | | Three-Phase Power Formula |

$$P = I \times E \times 1.732$$

$$P = 416 \text{ amperes} \times 208 \text{ volts} \times 1.732 = 149{,}866 \text{ VA}$$

$$\frac{149{,}866 \text{ VA}}{1000} = 149.8 \text{ kVA}$$

3. a	303	Table 430.91
4. d	304	430.102(B), Ex.(a)&(b)
5. a	559	695.3(A)(1)
	561	695.6(D)
6. d	411	517.17(A)
7. c	565	700.12(A)

120 volts × 87.5% = 105 volts

| 8. b | 286 | 430.6(A)(2) |
| | 293 | 430.32(C) |

54 amperes × 130% = 70.2 amperes

| 9. c | 344 | 500.6(B)(1) |
| 10. a | 359 | 502.15(2) |

ANSWER	**NEC PAGE #**	**REFERENCE**
11. d	162	314.16(B)
	162	314.16(B)(1),(2)&(5)
	163	Table 314.16(B)

(2) Size 6 AWG ungrnd. cond. = 2 × 5.00 cu. in. = 10 cu. in.

(1) Size 6 AWG grnd. cond. = 1 × 5.00 cu. in. = 5 cu. in.

(1) Size 8 AWG equip. grd. cond. = 1 × 3.00 cu. in. = 3 cu. in.

(2) Internal clamps = 1 × 5.00 cu. in. = 5 cu. in.

(1) Pigtail = 1 × 0.00 cu. in. = 0 cu. in.

TOTAL = 23 cu. in.

| 12. a | 290 | 430.22(E) |
| | 290 | Table 430.22(E) |

18 amperes × 85% = 15.3 amperes

*NOTE: When sizing conductors for intermittent duty motors, the NAMEPLATE rating shall be used.

13. c	273	424.39
14. c	194	354.20(B)
15. d	539	680.43(B)(1)(a)
16. a	172	324.10(B)(2)
17. a	440	530.51
18. a	269	422.48(A)
19. a	47	210.8(B)(1)

*NOTE: Article 210.11(C)(3) does NOT apply to hotels & motels.

20. b	444	547.8(C)
21. a	560	695.5(B)
22. a	576	725.26(B)(1)
23. d	454	550.31(1)
24. c	243	404.8(A)
25. a	349	501.10(B)(2)(2)
	359	502.10(A)(2)
26. d	443	547.5(A)

ANSWER	NEC PAGE #	REFERENCE
27. d	212	382.15
28. d	440	530.51 & .52
29. c	353	501.30(B), Ex.(1) & (2)
30. c	131	300.22(B)
31. a	166	314.27(B)
32. a	369	504.70(C)
33. d	72	230.2(A),(B)&(C)
34. a	61 61	220.56 Table 220.56

14.00 kW – range
5.00 kW – water heater
0.75 kW – mixer
2.50 kW – dishwasher
2.00 kW – booster heater
2.00 kW – broiler
—————————————
26.25 kW   total connected load

26.25 kW × 65% = 17.06 kW

*NOTE: The NEC states, however, that the demand load shall not be less than the two largest rated pieces of equipment.

14.00 kW + 5.00 kW = 19.00 kW demand load.

| 35. c | 250 | 408.35 |

$$\frac{100 \text{ lighting branch circuits}}{42 \text{ allowable circuits per panel}} = 2.3 \text{ or 3 panelboards}$$

36. a	111 112	250.122(A) Table 250.122(A)
37. d	333	480.9(A)
38. c		Three-Phase Power Formula

P = I × E × 1.732
P = 600 amperes × 208 volts × 1.732 = 216,154 VA

216,154 VA × 80% = 172,923 VA

ANSWER	**NEC PAGE #**	**REFERENCE**

39. b

320   450.3(B)
322   Table 450.3(B)
83   240.6(A)
        Three-Phase Current Formula $I = \dfrac{VA}{E \times 1.732}$

FOR PRIMARY:

$I = \dfrac{150,000 \text{ VA}}{480 \text{ volts} \times 1.732}$

$I = \dfrac{150,000}{831.36} = 180.43$ or 180 amperes

180 amperes $\times$ 250% = 450 amperes

FOR SECONDARY:

$I = \dfrac{150,000 \text{ VA}}{208 \text{ volts} \times 1.732}$

$I = \dfrac{150,000}{360.26} = 416.37$ or 416 amperes

416 amperes $\times$ 125% = 520 amperes

*NOTE: The next largest standard size overcurrent
 protection device is rated at 600 amperes.

40. c

516   645.7
130   300.21

41. c

530   680.8
530   Figure 680.8
531   Table 680.8

22½ ft. (minimum height above water)
2 ft. (diving board height above water)
22½ ft. − 2 ft. = 20½ ft.

42. d

47   210.8(B)(1)&(4)

43. a

135   310.10, FPN# 2

ANSWER	**NEC PAGE #**	**REFERENCE**
44. d	27	Article 100
	49	210.19(A)(1)
	58	Table 220.12
	59	220.14(K)(2)
	483	600.5(A)

6,000 sq. ft. $\times$ 3.5 VA = 21,000 VA
21,000 VA $\times$ 125% = 26,250 VA (lighting)
6,000 sq. ft. $\times$ 1.0 VA = 6,000 VA (receptacles)
26,250 VA + 6,000 VA = 32,250 VA (lighting and receptacles)

120 volts $\times$ 20 amperes = 2400 VA (one circuit)

$$\frac{32,250 \text{ VA}}{2400 \text{ VA}} = 13.4 \text{ or 14 circuits}$$

14 circuits + 1 sign circuit = 15 circuits total

45. c	361	502.115(B)
46. d	515	645.5(B)
47. b	351	501.15(C)(3)
48. a	57 & 58	220.12
	58	Table 220.12
	60	220.52(A)&(B)
	60	Table 220.42

4,000 sq. ft. + 2,000 sq. ft = 6,000 sq. ft.
6,000 sq. ft. $\times$ 3 VA = 18,000 VA
Three small appl. circuits at 1,500 each = 4,500 VA
One laundry circuit at 1,500 VA = 1,500 VA
18,000 VA + 4,500 VA + 1,500 VA
= 24,000 VA (total connected load)

First 3,000 VA at 100% = 3,000 VA
Remainder 21,000 VA at 35% = 7,350 VA
3,000 VA + 7,350 VA = 10,350 VA total demand load

49. b	311	Table 430.250
	561	695.6(C)(2)
	290	430.22(A)
	142	Table 310.16

FLC of 25 HP motor = 74.8 amperes $\times$ 125% = 93.5 amperes

*NOTE: Size 3 AWG at 75° C is rated at 100 amperes.

ANSWER	NEC PAGE #	REFERENCE
50. d	560 312 83	695.5(B) Table 430.251(B) 240.6(A)

*NOTE: The fire pump motor circuit protective device must be sized to carry indefinitely the locked-rotor current of the motor. According to Table 430.251(B), the locked-rotor current of the motor is 404 amperes. The next largest size overcurrent protective device has a rating of 450 amperes

| 51. c | 285
311
295
296 | 430.6(A)(1)
Table 430.250
430.52(C)(1)
Table 430.52 |

FLC of 10 HP motor = 30.8 amperes × 250% = 77 amperes

52. a	301	430.82(A)
53. b	249	408.3(E)
54. b	104	Table 250.66

500 kcmil × 4 conductors = 2000 kcmil total

*NOTE: Table 250.66 requires aluminum conductors of over 1750 kcmil to have a copper grounding electrode conductor of size 3/0 AWG.

55. c	258	410.30(C)(1)(2)a.
56. d	130	300.22(A)
57. a	142	Table 310.16

*NOTE: Size 250 kcmil 75° C rated conductor with an ampacity of 255 amperes is required.

255 amperes × .82 (temperature correction) = 209.1 amperes
OR 200 amperes ÷ .82 (temp. correction) = 243.9 amperes

58. b	37	110.26(E)
59. c	75	230.43(15)
60. b	104	250.66(A)

ANSWER		**NEC PAGE #**	**REFERENCE**
61.	b	140	310.15(B)(4)(a)&(c)
62.	d	254 & 255	410.4(E)
63.	a	96	250.24(A)(5) 408.3(C)
64.	c	429	520.41(A)
65.	b	401	515.8(A)
66.	c	83	240.6(A)
67.	a	38	110.31
68.	c	210	376.22
69.	d	210 632	376.22 Chapter 9, Table 5

4 in. x 4 in. = 16 sq. in. $\times$ 20% = 3.2 sq. in. (permitted fill)

$$\frac{3.2 \text{ sq. in. (permitted fill)}}{.3237 \text{ sq. in. (wire)}} = 9.8 \text{ or } 9 \text{ wires}$$

70.	a	73	230.7, Ex.1
71.	d	102	250.52(B)(2)
72.	d	72	230.3
73.	b	48	210.12(B)
74.	b	78	230.79(B)
75.	c	264 254 30	410.10(C)(2)&(9) 410.4(D) Article 100

ABOUT THE AUTHOR

H. Ray Holder of San Marcos, Texas, has worked in the electrical industry for over forty years as an apprentice, journeyman, master, field engineer, estimator, business manager, contractor, inspector, and instructor. Mr. Holder is a graduate of Texas State University and holds a Bachelor of Science Degree in Occupational Education. He was awarded a lifetime teaching certificate from the Texas Education Agency in the field of Vocational Education.

He is a certified instructor of electrical trades. His classes are presented in a simplified, easy-to-understand format for electricians.

He has taught thousands of students at Austin Community College at Austin, Texas, Odessa College at Odessa, Texas, Technical-Vocational Institute of Albuquerque, New Mexico, Howard College at San Angelo, Texas and in the public school systems in Fort Worth and San Antonio, Texas. He is currently Director of Education for Electrical Seminars, Inc. of San Marcos, Texas.

Mr. Holder is an active member of the National Fire Protection Association, International Association of Electrical Inspectors, and the International Brotherhood of Electrical Workers.